FARMING IN DORSET

Diary of James Warne, 1758
Letters of George Boswell, 1787–1805

FARMING IN DORSET

Diary of James Warne, 1758
Letters of George Boswell, 1787–1805

Edited by
J. F. James and J. H. Bettey

DORSET RECORD SOCIETY
VOLUME 13

© Dorset Record Society

Published 1993 Dorset Record Society
Dorset County Museum, Dorchester, Dorset

Designed by Laurence Keen
Typeset in Linotronic Baskerville by Character Graphics, Taunton, Somerset
and printed by Henry Ling Ltd., The Dorset Press, Dorchester
on Ganton Antique Wove 80 gsm vol 17.5.
Cover Lugo Grandee 216 gsm

British Library Cataloguing in Publication Data
Farming in Dorset: James Warne's Diary, 1758
and George Boswell's Letters, 1787-1805
I. James, Jude II. Bettey, J. H.
942.33073092

ISBN 0-900339-12-8

Cover illustration: from map of Winfrith Down by T. Sparrow, 1770
(Dorset Record Office D/WLC: P1/1). Reproduced by permission of W. Weld Esq.
Title page illustration by Thomas Bewick (1755-1828)

This facsimile hardback edition reproduces the 1993 text unchanged. It was published by
the Dorset Record Society in 2014 and printed in Great Britain by Lightning Source
ISBN 978-0-900339-19-6

CONTENTS

The Society is grateful to
The Marc Fitch Fund
Wessex Water plc
and the
Friends of Dorset's Archives
for their generous grants
towards the cost
of this volume
and to the County Archivist
Northumberland Record Office
for permission to publish Boswell's Letters

THE DIARY OF JAMES WARNE 1758

INTRODUCTION

The diary of James Warne, though covering only a single year, records remarkably fully the life, work and attitudes of a moderately well-to-do tenant farmer before agricultural 'improvement' had had any marked impact on the Wessex counties. Professor G.E. Mingay describes it as 'the most detailed and interesting document of its kind' he has seen.[1] The diary is supplemented by a large manuscript volume of 270 pages entitled *James Warne's Day Book*.[2] This valuable work was compiled by Warne between September and November 1798 using notes and material he had accumulated over several decades. It also provides a detailed account of a legal dispute from 1749 to 1753 over the use of water and watermeadows at Bovington. The plaintiff was the Revd. Thomas Hooper and the defendants Edward Weld, George Clavell and other proprietors. The account is composed as a long poem supplemented by explanatory footnotes.

It is evident that Warne was fascinated by genealogy, for his *Day Book* is particularly rich in information about himself, his family and his relations (see Appendix 2 for the Warne family tree). His own entry comprises a useful autobiographical sketch.

> WARNE, James (my Self, the Author of this Collection) was the Only Son of my Father, was Born at ARNWOOD in the parish of Hordle near Lymmington, Hants, on Sunday 17th April 1726, and at about 2 years of Age in May 1728 my Father removed from Arnwood Farm to Firwood[3] Farm, about 4 miles South of Cranbourn, Dorset. This Farm my Father Occupied 7 Years, but at the End of Six first Years my Father took an Estate at Northend in the Parish of Harbridge about 4 [miles] North of Ringwood, both in Hants. In May 1734, My Father placed

my Brother-in-law, Henry Loxly, and his Wife, letting them a small Dairy and He to superintend the Business there in my Father's absence and in May 1735 My Father's Farm at Firwood being expired The Family moved from Firwood to Harbridge where my Father dwelt till about the End of April 1743. Having had notice from James Willis, Esq., a Noted Attorney of Ringwood (his Landlord) that he intended to raise his Rent, or Else that he must quit the Estate, My Father choosed the latter, and in July 1742 (My Father having no Lease of Harbridge), He took Bovington of the Revd Mr Thomas Hooper, Clerk of Wimborne St Giles, for a Term of [*space*] Years. For the first [*space*] Years he gave £200 per Annum and for the last [*space*] Years the Annual rent of £210.

In his diary he makes a reference to his birthday under the date 28 April, noting that he was born on the 17 April (Old style). From 1752 he celebrated his birthday on 28 April to take account of the eleven days lost when the Gregorian Calendar was adopted on 2 September, 1752.

Joseph Warne, James's father, was a farmer of some substance, able to move from tenancy to tenancy to maximise his benefits. Thus he was able to enter into the Bovington Farm lease whilst still holding an unexpired term on Harbridge Farm. Bovington changed ownership during Joseph Warne's tenancy, being purchased by James Frampton of Moreton in 1757. Later, whilst still tenant at Bovington, Joseph Warne appears to have held the principal tenancy of both Woodstreet Farm, Wool (from May 1758), and Bindon Farm of Mr Edward Weld. His will reveals him as a man of substance, able to bequeath, in addition to a number of smaller bequests, 100 guineas to each of his nine grandchildren. His eldest grandson, Joseph Kingston Warne, was given the lease of Bovington Farm in October, 1773, some three years before his grandfather's death. This was the year in which Joseph Warne's second wife, Rebecca, died and probably indicates that he was no longer able to run the farm. Joseph Kingston Warne was bequeathed all the farming stock, all the implements of husbandry and 'All my Household Goods and Furniture, Beds, Bedsteads, Linnen, Woollen, Plate, China and all other goods of every sort and kind'. Joseph Warne died at Bovington Farm on 17 June, 1776.

James Warne was sixteen when his father took the tenancy of Bovington Farm and, although not specifically recorded, it is

8

quite evident that he worked with his father on the farm. In so doing he built up the knowledge and experience of farming that was to stand him in such good stead throughout his adult life.

At East Stoke church on Thursday, 21 November, 1751, whilst still living with his parents at Bovington, James Warne married Ann White, the eldest daughter of Stephen White, a tenant of Hethfelton Farm, East Stoke, about a mile and a half from Bovington. Ann had been born at Bockhampton on 4 November, 1727. Throughout the diary James referred to his wife as 'Nanny', and over a period of sixteen years she was to bear him ten children, six boys and four girls, the last when she was 38 years of age.

On 15 May, 1755, at the age of 29, he struck out on his own, taking the tenancy of Woodstreet Farm, Wool, part of the extensive estate of the Welds of Lulworth Castle. He remained at Woodstreet for three years and on relinquishing that farm took the tenancy of Turners Puddle Farm from Mr James Frampton of Moreton House. The diary clearly shows that he was actively engaged in farming Turners Puddle months prior to entering upon the tenancy and taking up residence. He remained at Turners Puddle for four years, and then leased Woodstreet Farm again, where he was to remain tenant for a further seventeen years. Following that, twenty years were spent as tenant of Bindon Farm, adjoining Woodstreet to the north, also a Weld property. In 1779 he leased Bagber Farm, lying in the south of Milton Abbas parish, for a short term until, finally, he entered the tenancy of his father's former farm at Bovington, taking over directly from his eldest son, Joseph Kingston, and remaining in occupation there for the rest of his days, dying in January, 1800,[4] at the age of 73.

Agricultural activities
Warne was farming in a world where many medieval agricultural practices were still in operation. The open field system was still widely practised in the area, but the Warne family were always tenants of the anciently enclosed or ring-fence farms. Woodstreet was already an independent farm at the time of the Domesday Survey in 1086. Turners Puddle Farm, which James Frampton had bought in 1752 for £2,100, was the largest of the three enclosed farms making up the parish of Turners Puddle.[5] The agricultural revolution had hardly started in the 1750s and Warne's diary gives no indication of

9

novel advances in agricultural methods. He lived in a world of traditional husbandry, and his values in relation to farming practices and his dealings with fellow farmers, his landlords and his labourers are clearly spelled out in his diary entries. It is evident he worked a long day, but it included what today we might term a 'social' element: he often had dinner with the farmers he visited, and a good deal of lending of equipment and stock took place between farmers.

As a working farmer James Warne was involved in day-to-day operations. He helped his men with the winnowing, worked with the women in the fields and engaged in heavy work such as carting and building a bank (16 October). However, he frequently worked only part of the day on the manual tasks and then rode off to market or to make a visit to his father at Bovington or to other farmers on business. Only once does he make reference to his regular daily tasks when he states, 'I Pitched the Fold as I do most Mornings' (24 August). From time to time he had problems with his workfolk, and this obliged him to undertake tasks which had been allotted to them. On 13 December neither of his labourers, George Sinnick (who was in any case under notice to quit) or Tom Hunt, would go to cart mud from the river. 'So I and dum Jack had down the 4 Horses and the 2 Carts and Jack and Old T. Hunt filled and I a little and the Dairyman drove it.' In fact the diary proves very revealing in the matter of relationships between Warne and his labourers. On 25 June Warne's farm servant Robin Locass went home without permission and on his returning Warne whipped him. The next day Locass felt so 'affronted' by his treatment on the previous day that he asked for his pay so that he could leave, as 'he would not bide to be Beat'. On Warne's reprimanding him, 'He took it into Consideration and went to Cart as usual'.

It is not, of course, only the problems that are of interest but also the means of engaging labourers. Sometimes individuals seeking work merely came and asked, as on 12 February or on 17 December. When Warne was visiting Salisbury on 2 October he was approached by a 19-year-old lad from Portesham offering his services. Warne evidently found the young man, John Antle, to his liking and asked him to call at Turners Puddle after he returned from Weyhill Fair, where he had been driving sheep. Antle arrived on 9 October and was engaged by Warne at 4 guineas (£4 4s), but as the month progressed the choice proved not to have been a wise one. On 18 February Warne agreed with his servant Daniel Dobb that he should serve a further year from the following 5 April for £4 plus a pair of gloves at harvest. It also appears from the diary that others were sometimes dissatisfied with the labourers, notably Mr James Frampton, who, Warne records on 15 June, 'found fault with the People that they did not Work enough', and on the 20 June he 'curses the Workman for a Parcel of Lazy Rogues'.

The diary provides a vivid insight into the general conduct of farming, as when in February we are informed that ridges are being set out in Lower Hog Lease in preparation for sowing. 36 ridges, each of half an acre, were laid out in eight days. These ridges assisted drainage in fields by creating gentle cambers.

The business of ploughing, sowing, harrowing and wrestering are detailed. In March horses were employed on fifteen days in these activities in Long Craft, Mountfield and Buckshill. The pace of activity increased in April when Five Acres, Pit Close and Hog Lease were ploughed in readiness for sowing. Oats and peas were sown in Buckshill, peas in Five Acres and barley in Pit Close and Hog Lease. The sowing of the winter wheat took place in November and we are informed that the field 'Cockroad' was planted with 39 bushels and 'Buckshill 6 Acres' with 19 bushels (18 November). It appears that between five and seven bushels of wheat were sown in a single day but greater quantities of barley, e.g. 12 bushels on 24 April. This work was interspersed with the business of threshing and winnowing the previous year's crops, which continued into late April, and with carrying out general maintenance on the farm, such as clearing ground, making up roads and tracks, repairing cow pens, collecting frith, driving animals, selling and buying produce at market, commencing work on new channels for the watermeadows and several other activities. In 1758 harvesting started on 14 August when Warne with 5 men began the wheat reap in Saltfield. Once the corn harvest was gathered, the first crop of apples was ready to gather for the autumn cider-making, which Warne describes in detail (e.g. 13 October).

His farming tools, as recorded in the diary, comprised suls and ploughs, harrows, drags and wrests, a winnowing fan and a cider press. Although sowing is frequently mentioned, there is no reference to a drill and it can be assumed that the usual practice of broadcast sowing was operated by Warne. He had carts and two wagons which performed the essential transportation services for his operations. His livestock consisted of cattle, sheep and pigs and, whilst at Woodstreet he had seven working horses, he increased this number to nine at Turners Puddle.

The Dorset system of renting the milking herds to a dairyman at the rate of about £3 per cow is well described (e.g. 4 March).[6]

Watermeadows had long been established in the Frome and Piddle valleys, and at Turners Puddle Warne was actively engaged in improving the watermeadow system, as many entries attest; he also attempted to improve drainage in the traditional way of ditching (e.g. 16 March).[7]

The diary reveals that Warne practised some diversification, for the entry on 20 November shows him to be systematically planting fir trees at Bovington as a commercial crop.

Role in the community
James Warne, like so many educated farmers, was to take an active role in his community. He was elected an overseer of the poor, served as parish clerk and acted in other capacities in the parishes of Wool and Turners Puddle, and was elected overseer of the poor for Turners Puddle on 31 March, before taking up full-time residence in the parish. James Warne's eldest son, Joseph Kingston Warne, after a six-year tenancy at Bovington, became tenant of Moreton Farm in about 1779, and whilst there he became a man of considerable stature in the local community. The last son born to him and his wife Leah (Biles) was Charles Warne,[8] an exact contemporary and personal friend of William Barnes, who was to become a famous Dorset antiquary, and author of *Ancient Dorset* (Bournemouth 1872).

Like another famous diarist, the Revd. James Woodforde, Warne rarely notices events outside his locality and yet it is obvious from his *Day Book* that he had a wide range of interests and concerns. On 22 February he noted, 'Good news (if True) Admiral Cole's has Taken 9 Of the St Domingo Fleet and its convey and an English Prisnor made his Escape from the French at Vigo'. And he notes a 'General Thanksgiving' held on Sunday, 3 September, for the success of English arms in capturing Louisburg on 24 July, 1758.

Travelling and visiting
The diary demonstrates very clearly the amount of journeying that Warne did on a day-to-day basis. Between January and mid-May, he rode regularly from Woodstreet to Turners Puddle, frequently calling *en route* on his parents at Bovington, or his in-laws at Hethfelton. The diary entry for Wednesday, 15 February, provides an insight into a typical day's activities.

Frequent journeys were made to the market towns on market days. Wareham, the nearest to Warne's farms, was visited on most Saturdays. He also made fairly regular excursions to Blandford on the same days and, less frequently, to Ringwood. The market day visits were essentially geared to the farming business but were not without social intercourse (e.g. 25 February).

In his journeyings one jaunt is of particular interest. On the morning of Saturday, 30 September, Warne set out with his wife, Nanny, and his sister-in-law Patty, for the village of Martin, where he had relatives. They broke the journey at Blandford before moving on to Martin. On arrival there James and Nanny Warne lodged with his brother-in-law, Henry Loxly, whilst Patty stayed with another brother-in-law, Brown. Next day they visited St Mary's, Fordingbridge, where Warne was impressed with the church band.

On the Monday they were to have started on the return journey, but as Sister Patty's horse was ill this had to be postponed. Warne, never one to lose an opportunity, then rode to Salisbury to call on Mr Gustavus Brander but as he was away he called instead at his mother-in-law's, Mrs Shaw, who lived in the Close. Warne then went to evening prayer in the cathedral.

Next day Warne, Nanny and Patty set out for Ringwood but interrupted their journey, first at Harbridge, where they stayed about an hour with another relation, Michael Saunders, and then at the tiny hamlet of Kent, where they visited Farmer Biles. From there they crossed the Avon to Ibsley. The final visit was to Farmer Mist at Moyles Court. The night was spent at Cousin Joseph

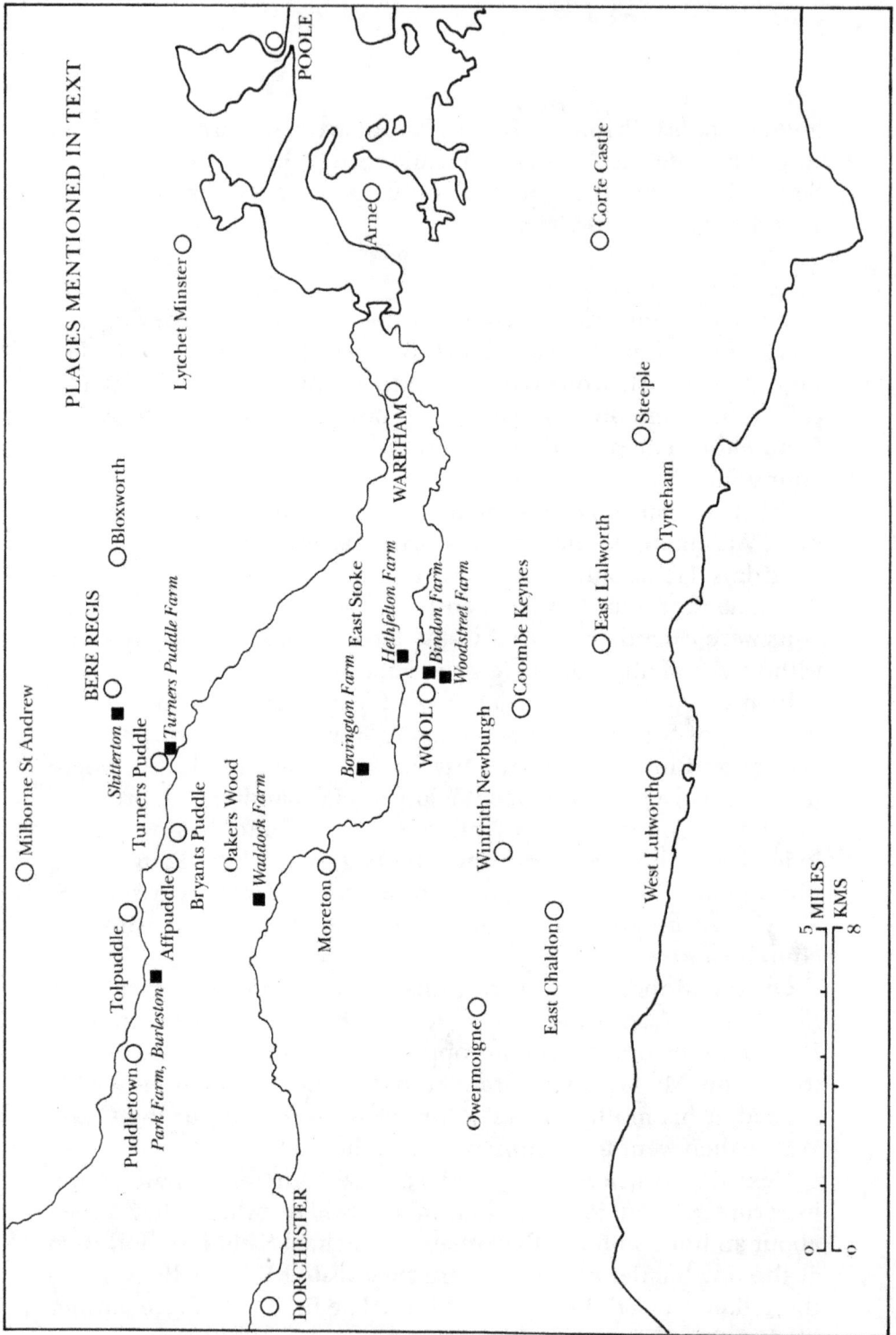

PLACES MENTIONED IN TEXT

Milborne St Andrew

BERE REGIS

Bloxworth

Lytchet Minster

Shitterton

Turners Puddle

Turners Puddle Farm

POOLE

Tolpuddle

Puddletown

Park Farm, Burleston

Affpuddle

Bryants Puddle

Oakers Wood

Waddock Farm

Arne

Moreton

Bovington Farm

East Stoke

Hethfelton Farm

WAREHAM

Owermoigne

Winfrith Newburgh

WOOL

Bindon Farm

Woodstreet Farm

Coombe Keynes

Corfe Castle

Steeple

East Chaldon

Tyneham

DORCHESTER

West Lulworth

East Lulworth

5 MILES

8 KMS

14

Warne's at Kingston, on the southern edge of Ringwood parish. Breakfast was taken at the home of some cousins called Jessop and a final visit made to Mr Willis's house. At 3 o'clock in the afternoon they set out on the return journey, arriving back at Turners Puddle between 8 and 9. All this was done on horseback over unturnpiked roads.

On one occasion, on 11 December, in order to get to Ringwood fair in good time for business, Warne set out from Turners Puddle at 4 o'clock in the morning, arriving at Ringwood between 8 and 9 just about half an hour after sunrise.

Another revelation is provided by the details of a funeral journey made on Sunday, 2 July. Mr Henry Lockyer, aged 63 years, bailiff to Mr Frampton, died suddenly on 29 June and his body was to be taken from Moreton to Hampreston, the family's home parish, for burial, a distance of about 20 road miles. The party of twenty-five men mounted on twenty-five horses set out at 8 o'clock in the morning and, reaching the World's End inn, were joined by other mourners. The party arrived at Hampreston at 3 in the afternoon and the interment took place at 4. James Warne was one of the six farmers who bore the funeral pall. On the return journey a visit was made to a Farmer Hayward of Wimborne, where their horses were fed and watered, and Warne arrived back at Turners Puddle between 9 and 10 in the evening.

Some journeys were purely for pleasure, interest and enlightenment, as when James and Nanny Warne set out for Weymouth to see the 'Great Fleet' (28 August).

The diary contains many other examples of journeys, mostly of a regular nature, which invoke a strong impression of almost constant movement. Even in churchgoing we find Warne, his family and other farmers frequently conducting a kind of circuit of the local places of worship.

Building at Turners Puddle farmhouse

The diary throws considerable light on building practice of the time, for, prior to moving into Turners Puddle farmhouse, Warne had requested structural improvements to be carried out. Mr James Frampton, his landlord, agreed to these proposals, and on 3 March he obtained the services of a well-known Blandford architect, Francis Cartwright, to make plans for the alterations.

Cartwright visited the farmhouse on 17 April. However, James Warne evidently fancied that he could make a better job than the professional architect and drew up his own plans for the modification to the house, recording in his diary that the masons would rather build to his plan than Cartwright's! (10 May) He actually showed his plans to the architect. The building activity caused considerable domestic upheaval as it progressed, but on 19 August he could record with satisfaction that most of the work was finished, leaving only a few tasks finally to complete the work.

Religious life

Warne was a devout Anglican and frequently attended two church services on Sundays. He was in the choir, which benefited from training by Elias Hibbs, a music tutor (e.g. 29 January). Warne notes every Sunday the text chosen for the sermons and gives details of the Psalms that were sung. Professor Derek Beales comments that the music of Warne's choirs fits into a widespread rural phenomenon in the eighteenth century; they must have been using a basic book of metrical psalms with a supplement or additional set of tunes, although the actual collection has not yet been identified. Psalms at this time had not given way to hymns in the liturgy of the Established Church, though they were widely sung by dissenters. Only once, on Christmas Day, does Warne use the word hymn, which undoubtedly refers to Nahum Tate's carol *While Shepherds watched their Flocks by Night*.

Warne also appears to have attended dissenters' meetings at Bere Regis (e.g. 13 August). However, he seems to have been dissuaded by the Revd. Mr Fisher, who spoke to him on one occasion about the spiritual errors of dissent (10 September). Nanny apparently was not so convinced, for on Sunday 15 October Warne remarks, 'Nanny rode to Beer to go to Meeting there.' Previously she had indicated that she did not like the sermons delivered at the Church of England services (2 and 23 July) and seemed to prefer the meetings. James, although prepared to attend meetings, did so in addition to the normal church services. Warne never states which sect is promoting the meetings, but in the context of his religious conviction it seems most likely that it was the Methodists.[9]

REFERENCES

1 G.E. Mingay, 'The Diary of James Warne, 1758,' *Agricultural History Review*, 38, part 1 (1990), 72.

2 DRO D. 527/1 (pages 21-30 inclusive torn out).

3. Firwood = Verwood.

4 Wool parish register [DRO PE/WOO: RE1/1], 1800: 'Mr James Warne, senior buried January 10th. Affidavit made'.

5 Information on Frampton's farms abstracted from DRO D/FRA: E68.

6 The system is described in J. Claridge, *General View of the Agriculture of Dorset* (1793), 14-15.

7 For background see B.J. Whitehead, 'The Management and Land-use of Water Meadows in the Frome Valley, Dorset', *Proceedings of the Dorset Natural History and Archaeological Society*, 89 (1967), 257-281; also G. Boswell, *A Treatise on Watering Meadows* (2nd edition, 1790).

8 '1801 August 9th. Charles, son of Mr Joseph Kingston Warne and Leah his wife was baptized' There had been an earlier son named Charles (baptised 11 March, 1794) but he had died in infancy [DRO PE/MTN: RE1/2]. Charles Warne died 1887.

9 As Dr Biggs states in *The Wesleys and Early Dorset Methodists* (1987), 58, 'Early Methodists did not regard themselves as Dissenters, but true Anglicans, and often avoided the label'.

THE MANUSCRIPT

The diary is printed and bears the title:

The DAILY JOURNAL :
OR, The
Gentleman's and Tradesman's
Complete Annual Accompt Book
For the POCKET or DESK
For the Year of our LORD 1758

It is leather bound on board and measures six inches (152mm) long by three and three-quarter inches wide (95mm) and is three-quarters of an inch thick (19mm). The diary is also an almanac containing a variety of information printed in eleven pages in the front and eighty-four pages at the back. The entry pages are divided horizontally across the open format into seven intervals for the days of the week starting on Monday. There are vertical divisions headed *Account of Monies; Rec'd £. s. d.; Paid £. s. d.; Appointments and Memorandums, or Observations.* Only for the initial entries does Warne follow these printed divisions; once in his stride he runs his daily record straight across the columns. The diary is now housed in the Dorset County Record Office (reference: D. 406/1). The weekday entries are written in brown ink and the Sunday ones, with a few exceptions, in red.

Measurements

It has to be noted that there were local variations of volumes and weights. The standard measurement for grain was the Quarter (Q. or Qtr.), which comprised 8 Bushels (B. or Bs.), each bushel comprising 4 Pecks (Pk. or Pks.). There were normally 8 gallons in one bushel. Pint measures were used for beans and peas (e.g. 22 April). A Hogshead was 63 gallons (1 April, 28 April). A Score was either 20 or 21 lbs (e.g. 21 July). Wool was measured in weights of $6\frac{1}{2}$ tods, 2 weights made 1 sack (18, 21 September). Fleeces varied in weight but were usually around $2\frac{1}{2}$ lbs each, 2-3 fleeces made a clove of 7 lbs., 4 cloves to 1 tod. Claridge gives 31 lbs to the wey/weight (J. Claridge, *General View of the Agriculture of Dorset, p. 9*).

EDITORIAL PRACTICE

The diary entries are often extremely cramped and difficult to read, as can be seen from the facsimile double page printed as a sample in Appendix 1. Warne makes extensive use of abbreviations and contractions and often omits words. His spelling is variable.

In editing the text, the aim has been to render the text intelligible to the general reader while still retaining some flavour of the original. Thus all lower and upper case letters have been copied as written by Warne, except that a full stop is always followed by a capital. Certain contractions used regularly have been spelled as we would today, since Warne's use is variable. Standard abbreviations have been silently expanded. So have the idiosyncratic abbreviations used regularly by Warne. These include: A.N. (Afternoon), F.N. (Forenoon), S.R. (Sunrise), S.S. (Sunset), U. (Uncle), X (Cross), O. (Old), 1.pt (last part), f.pt (first part) and Ch. (Church), 2 (to). The diarist frequently uses only initial capitals to denote personal names, and in most cases the full name has been completed in square brackets.

Warne's punctuation has been followed unless it has been thought to be misleading. In some cases it has been necessary to judge where sentences start and finish and punctuation has been amended accordingly. Warne's use of apostrophes is haphazard, and therefore these have been added where it is clear that modern convention calls for it.

Editorial insertion has been kept to a minimum, though where it occurs it is indicated by the presence of square brackets. Trivial errors have been corrected without comment. There remain some code or shorthand characters that have defied analysis; their presence has been remarked on.

ACKNOWLEDGEMENTS

I would like to thank the Dorset Record Society for allowing me to edit this exceptional diary. It was first brought to my attention in 1977 by Miss Margaret Holmes, then Dorset County Archivist, when I was preparing a dissertation on Dorset agriculture; I am grateful to her not only for this but also for much help and advice. Subsequently, the present County Archivist, Mr Hugh Jaques, has been generous with his assistance, and Miss Sarah Bridges, the Assistant County Archivist, has used her great skill and experience to sort out the difficulties of Warne's text. She has rescued me from many mistakes.

The Council of *The Hatcher Review* provided me with the first public forum for a lecture on Warne's Diary, and several members of the audience offered additional information, which was later incorporated into the published version (J. James, 'A glimpse of an old world: the diary of James Warne for the year 1758', *The Hatcher Review*, 4, No. 33 (Spring 1992), 30-49). Much of the introduction to the present volume is based on material which first appeared there in an expanded form, and I am grateful to the editor for permission to use it.

My thanks are due to Miss B. F. M. Frampton of Moreton House for making available to me her family's archives; to Sir Gilbert and Lady Debenham for permission to visit their house in Turnerspuddle, which once was lived in by Warne; to Mr Roger

Peers, former Curator of the Dorset County Museum, for drawing my attention to the Warne family portraits; to Mr Ted Baker of Ringwood for information on Warne's family connections there; to Dr Brian Tippett of King Alfred's College for guidance on Warne's poetic ability and for a copy of *The Art of English Poetry*; and to the staff of the Perkins' Agricultural Library in the University of Southampton, who have been unfailingly helpful.

I owe a special debt of gratitude to Mr Laurence Keen, Mr George Clarke, Mr Roger Peers and Miss Sarah Bridges, who have undertaken the task of seeing the volume through the press.

Finally, I would like to express my thanks to my wife, whose perceptive comments and advice throughout my work on Warne's Diary have been of inestimable value.

J.J.

THE DIARY

---

1 January [Sunday].   Hard frost Continues.
                    [shorthand characters] £93 . 17s. 0½d.
I Went to Coomb Church in Fore noon Where was but a Small
Congregation, so there was only Prayer which began about 11
o'clock and After Prayer The Sacrament. I and [my] Wife went to
Wool Church in After Noon, where Mr Bank[e]s of East Lulworth
preached his Text 1 Corinthians 13.12 and We Sung Psalms 96ª.
100, 48ª/34, etc.

2 January.   Carried the first Load of Wares to Piddle, viz: a 2
H[atched] Ware and Sluice into Lillington's Mead by Fords. And
it being hard frost the Horses came Home that Night.

3 January.   Rain in the Night and Frost went Quite off. Horses only
Bedded up the Barton. T. Coffin Seized John Lambert's Goods for
House rent and William Snell lent him 21s. to redeem it.

4 January.   To Mr Weld rent to last Michaelmas in Cash £49. 14s.
od. of Mr W[eld] for Disburstments £10. 6s. od. *[The sums are
crossed through.]* Horses carried abroad Earth and Sand in the
Moor. I rode with my Father and dined at Farmer Garland's at
Lodge [Farm] and Afterward went to Lulworth Castle and Paid
Mr Weld my Rent.

5 January.   To Dine at Mr Robinson's at Stoke. But received an
Account of the Death of Brother Nathaniel Langford (who Died
at Portsmouth, December 27th) so did not Dine at Stoke, But
went to Heffelton to Condole with Mother White.

6 January.   The Horses went to Piddle with both Dung carts, and Carried Wheat, Barley, Dust and Oats, and 1 Cart Carried Stiles etc. from Bovington. J[ohn] Allen began setting the Wares at Piddle. Killed the Last of the Two Fat Pigs which weighed about 18 score. Between October 10th last and this time Ground at Turners piddle Mill:

8Q 1B 2Pk	of Barly Whereof Sold	3Q 3B 1Pk	
3Q 7B 1Pk	and gave the Barton Pigs	0Q 4B 0Pk	
4Q 2B 1Pk	So much the 2 pigs Eat	3Q 7B 1Pk	

7 January.   J. A[llen] and Simon Hallet with Labourers about the Wares at Piddle. Very Mild Pleasent Weather. Horses carried Earth out of Heath into 5 Acres at Bucks hill.

8 January [Sunday].   I Went to Stoke Church. Text John 8. 51. Mild Weather. Farmer White returned from Portsmouth this Evening.
[No entries 9 to 24 January inclusive]

25 January.   Eleven of us Singers with Farmer Sexey and John Brown, spent an Evening together at the Rose and Crown. The Company very sober.

26 January.   Stephen White Warne Born January 26th 1754.
[No entry for 27 January]

28 January.   My Horses and Farmer Talbot's carried 2 Load of Frith from Bovington to Piddle Farm, and in the Evening William Locass and W[illiam] Hunt came with the Wagon and 5 Horses to Woodstreet brought Home W[heat] Meal.

29 January [Sunday].   Went to Wooll Church. Text 20 Leviticus 26 and We Sung Psalms 84, 149[a], 21 and Stayed and Sung after Prayer with E. Hibbs our Tutor. Close Cloudy Weather.
[No entries for 30, 31 January and 1 February]

2 February.   Let to John Rawls at Puddle the Hedge at the West Side of Lower Hogs lease, to make it at 2d. 3q[?3 quarters] a lug at 16 feet ½ to the lug to Throw a Ditch a good spit deep, and well shovelled, and 3 Spit Wide. Simon Grant Buried at Winfrith. Carried 2 Load of Frith From Bovington to Piddle into the Orchard at Fords.

3 February. I, Simon Hallet, and Thomas Talbot began parting the 4 Cow pens by Brick Barn. They began Sawing out the Posts last Tuesday. J.R. Standley and Thomas Vocam measured the Lot. T[homas] Vocam routed in Hazzle Wood (which is the 5th lot from Honix, and first finish) and it is 154 Lug and ½. Nanny at Wool saw Mr Bewnel and his 2 Sisters who promised her to come and see us next Monday.

4 February. At Mrs Pearce's at Beer.
<table>
<tr><td>Bought Figs</td><td>1s. 2d.</td></tr>
<tr><td>To Collarmaker Whindle</td><td>2s. 0d.</td></tr>
<tr><td>To Mrs Harding for Candles</td><td>6s. 8d.</td></tr>
<tr><td>To John Phillips for Driving Plow</td><td>2s. 6d.</td></tr>
</table>
I and Daniel Cobb Surveyed the Farm Orchard Home Plot, Church and Churchyard and Bartons at Turners piddle and in the Evening I began Planning it.

5 February [Sunday]. Went to Stoke Church. Text Deuteronomy 16.19. They Sung Psalms 139, 84ᵃ, 100. Mr Vanderplank and Mr Bartlet of Beer came to Visit us in the Evening.

6 February. Mr Vanderplank and Mr Bartlet Departed about Sunsett. To Mrs Hayte pipes and Tabaco 0. 0. 41 *[sic]*. Mr Bewnell and his Sisters to come. He came. But not his sisters. I must go to Piddle to Set Out Ridges in lower hog lease. I Went, and struck out 9. Called at Bovington as I came back. A little rain in forenoon.

7 February. I Struck out 17 more Ridges at Piddle and brought home Some Barley and Wheat Meal. Moist Air, a little rain in the Evening.

8 February. To Mr Bewnel of Coomb for 18 young Cattle running the Summer in Coomb heath 18s. 0d. and 12 Poor Rates 8s. 0d. I rode to Bovington and from thence with my Father to Piddle

and saw some particulars there brought home W[heat] Meal and
Went to Coomb and spent the Evening and Mr Bewnel. Very drying Wind at North West.

9 February. Of Alexander Grant (by the hands of his Daughter,
Ursula Brown) 3 Poor rates 14s. od. 5 of my horses carried a
Load of Hay from Turnerspidle to Grange for Farmer Talbot and
came back and laid here. The Farmer came this Way and Told me
He paid T. Hunt 16s. 6d. for Settling the Water the first year
which at 6d an Acre is 33 Acres.

10 February. William Locass went to Piddle with the wagon and
carried dust, Oats, Wheat, and Barley and some Thorns on Top. I
and Nanny rode to Heffelton and There Dined with Mr Brine of
Chettle Down.

11 February. Paid T. Hunt for Hedging and Setting Wears   o. 19s. 6d.
               Of him for Wheat   o. 11s. 3d.
               To William Hunt for Plowing   o. 12s. od.
               Of him for Wheat   o. 9s. 7d.
               To John Philips at Plow   o. 1s. 3d.
William L[ocass] *[shorthand characters in red ink]* Wagon carried a
Load of Frith from Bovington to Piddle. Paid Thomas Hunt,
William Hunt, and John Philips at Piddle. I gave William Locass
the Offer of my Dairy of 12 Cows at Piddle which He is to consid-
er Off.

12 February [Sunday]. I and Wife went to Wool Church in
Afternoon. Text Romans 12.21 first part and We Sung Psalms 66,
132ᵃ, 29/16ᵃ, 26ᵃ, 68, etc. A Man from Bryants piddle came to
Ask for Mowing! Farmer Willshear of Grange Drove by here 2
Cows and Calves for Dorchester fair Tomorrow! See how the
Sabbath is regarded!

13 February. P. Burden not at Work He having a Bad Thumb and
J. Lambert went home again. He struck himself in the face last
Saturday. A Brisk Drying Wind. My Father called here about
Noon. Afterward I Waited on Mr Bewnel, and we Appointed next
Thursday as Under. I and D. Cobb Mended the Bull's pound.

14 February.   To Mrs Elliott for Malt etc. 0. 11s. 4d. P. B[urden] and J. L[ambert] both at work this day. I Surveyed the Reek yard and South view of Piddle House. William Locass don't choos to take a dairy. I Rode to Bovington and Piddle and made an End Stricking out of Ridges in Lower Hogs lease, which is Called 18 acres and there are 36 ridges in it which is just 2 to an Acre. A little Frost in the Morning, rain all the Afternoon.

15 February.   I cleaned the Cowpens etc. Walked round the fields and into Hazzle Wood. I began a Draught of Piddle Rick yard. Father and Mr White came and their Dairyman, Gold Grant. Drying wind with Showers. Horses ploughed with both Sulls as yesterday in Hoglease. Farmer Sabbin came to Bovington last Night and to day Father Rode with him to See Mr Gostelowe's Farm at Chaldon. Coming back they drank Drams so free at Mr Filton's that the Farmer fell from his horse at Hyford Gate, a Great mercy He was not drowned!

16 February.   I Let to Philip Whittle of Clift a Dairy of 20 Cows at Turners Puddle, To be made at Fords To have £3 a Cow. To Summer Feed, Fords 6 Acres, Alderham's and Frogham, Springs and Brockhill 10 Acres and Rack Mead, Lillington's Mead and Cicilias for Yea grass. To meet Mr Bewnel at Wooll at 7 in the Morning. To go and see Dairy Cows at Tomson. Mr Bewnel Sent word He could not go to day but would any day next week, so I rode to Bovington and let my Father know it. Farmer Sabbin departed as I rode to Piddle. I measured the Barns and Bartons.

17 February.   *[Friday: this entry in red ink]* Fast day. Dry Windy, Sun Shine Weather. I went to Stoke Church at 10. Nanny, Father and son, J[oseph] K[ingston], at Wooll [12.15], Coomb [2.30]. Tex[t] at all 3 James 4. 8, 9, 10, sung at Stoke Psalms 16, 18, 28. We Sung at Wooll Psalms 16[a], 68/35, 26[a], and From Wool went to Coomb Church with Mr Robinson and There Sung Psalms 48[a], 96[a], 26[a], 16[a].

18 February.   I Rode to Bovington and from thence to Wareham Market, the 1st Time Since October 3rd. I Sold Some Pease there for 5s 8d. a Bushel delivered here at home. Barly was 28s [a]

Quarter, Wheat 6s. 6d. and 6s. 8d. [a bushel]. To Mr Chisman House rent for T. Barns £1. 3s. 9d. To Dine with Brother Florence and Company next Tuesday at Heffelton. Agreed with Daniel Cobb to Serve me another year from April 5th next for £4 and [a] pair of Gloves at Harvest.

19 February [Sunday]. It rained all Day long. I Went to Stoke Church in Forenoon. Text Hebrews 12. 3 and they sung Psalms 92, 34 2nd part. *[space]* Florence etc. of Sturminster Marshall called here about Sunset To enquire the Way to Bradle in Purbeck.

20 February. Intended to have rode to Tomson with Mr Bewnel and Father to See Dairy Cows, but rained so in the Morning Mr B[ewnel] did not set out. I was at Bovington and came from thence with Father and James Keat of Corfe [Castle] and look out some stone for Cow pen posts at Bindon and they both Dined here with Mr Chisman of Arne and all departed about sunset. Set out the Hedge between Long Close and Bottom.

21 February. I rode to Piddle Early and gave Orders and came back and J. Daniel, P. B[urden], D. C[obb], J. L[ambert] and Father's 2 Wagons, 6 horses, 2 Men, 2 Boys and 3 Labourers took in the rest of my Barly Reek which pretty well filled the Barn. We Killed 8 Rats. Nanny rode to Heffelton to Dinner. I rode up and came home with her in the Evening. Brother F[lorence] was not there but there was Sister and Mr Brine, and Mrs Bartlet, Mr Bartlet and John Burleigh Dugdale was there at Dinner but gone. A good dry day but brisk Wind.

22 February. Rain in the Night and Morning. Cut down Barly Mow etc. Rode to Piddle after dinner. Good news (if True) Admiral Cole's has Taken 9 Of the St Domingo Fleet and its convey and an English Prisnor made his Escape from the French at Vigo. D[aniel] C[obb] *[shorthand characters in red ink].* Horses Earth cart at Buckshill and Ordered them to come home tomorrow.

23 February. I rode to Piddle and Shitterton before Sun rising, came back by Bovington. Father went yesterday to Piddletown on a Commission between Mr Hayte and Mr Alner and was not come

26

home. My wagon came home and carried Dust, 12 Bushels Oats and 36 Bushels Wheat. Joseph Alner called here and wanted Father with Farmer Garland to go tomorrow and decide a Case between him and Farmer Chip of Beer. Thomas Coffin came this Evening and I wrote him an account of his horse that was stole January 26 and found February 12th. Joseph Davison Called as He was going from Piddletown to Highwood and brought word from my Father that I should ride with Farmer Garland to Bovington and If Father was not at Home to Beer with Farmer Garland in his Steed.

24 February.   I rode to Bovington and there met Farmer Garland (Father was not come home from Piddletown). We rode to Turnerspuddle and called Joseph Alner and rode with him to Beer and There met us Farmer Vicar of Shapwick. Farmer Chip's Other Refferee did not come nor would he sign Bonds of Award which Joseph Alner there had and Offered to do if Farmer Chip would so we agreed on Nothing. Got home between 7 and 8 Clock.

25 February.   I rode to Bovington in the Morning with the Butter and from thence to Wareham Market. Mr Balson of Martins Town and Farmer William Style of Forthington called at Bovington in the Morning and rode to Wareham Market with my Father. I came home with my Father into Woolbridge lane. Uncle John White was here and spent the Evening till about 10.

   To Mr Clench for worsted 10 Oz   0. 2s. 6d.
   To Dr Seward for my Son Robert Duel 0. 3s. 0d.

26 February [Sunday].   Went to Stoke Church in Forenoon. Text 1 Corinthians 4.18 and they sung Psalms 149^a, 71, 145^a. Blessed Mild Morning. I and Nanny went to Wool Church in Afternoon. Text 1 Corinthians 4.18. We sung Psalms 108, 48^a, 117^a. Staid and sung after Prayer with our Tutor.

27 February.   Horses restered and carried Earth. Rode Over to Piddle, Nanny rode with me to Wool and so to Bovington. I called her as I came back and came home together. Father was here in the Morning and rode from hence to Moreton and there Met

Farmer Simon Elliott and Valued some Meadground between Mr Frampton and James Drake. The Chimney at Piddle Mill fell down this Day. A Drying Wind.

28 February. My Wagon carried a Load of Wares from Bovington to Piddle and Afterward carried gravel and 3 [horses] Restered. I and D[aniel] Cobb parted part of the Cart house for the Weanling Calves. About Noon Father brought up Mother and left here here. In the Evening did a little of my Plan of Piddle.

1 March. Rain till Noon. I rode to Bovington and so to Piddle. I was very wet, horses not out till After I got there, then Bedded the Barton at Farm. I surveyed the Barns, Cart house and dwelling house, and laid out a place for the New Barn. John Sillaway Fetched Mother. I got home [between] 7 [and] 8. More rain.

2 March. From Piddle my Father and I rode and saw Mr Bewnel's dairy Cows at Thompson but did not deal, but I drank to much Cider so that I was very sick. Got home about 8. To Meet Mr Frampton at Piddle at 10 in the Morning to fix a Place to build the Barn in. Met there and fixed on a place and Mr F[rampton] resolved to pull down the Kitchin and some more and build two good rooms with Chambers and Garrets Over. Horses at Gravel cart and restering in Long Craft.

3 March. I rode to Bovington and there surveyed the Wheat Barn and am to draw a Plan thereof for Mr Frampton and from thence to Piddle. Saw my work folk and looked on Robert Philip's Bargain, which I think I shall Take. Wagon carried 2 Load of Wares to Piddle. After I came home went to Mr Bewnel's at Coomb and talked further about his Cows but did not Deal. Mr Bestland of Wareham called to Buy Pease. He bid me 5s. 8d. a bushel and would fetch them but I Asked him 6s. and He did not choose to give it.

4 March. To Mr Bewnel Earnest for 26 diary Cows   £10. 10s.
      he is to have more at the delivery   £119. 18s.
which will be May 13th or 15th which sums make in the whole £130. 8s.

4 horses Harrowed resters and plowed in Long Craft. Mr Bewnel called here and We Agreed as in t'other leafe *[i.e. entry on opposite page]*. I rode to Wareham Market, Saw Father there [and] showed him what we had done and He let Bovington dairy to Old Haines, Mr Bewnel's dairy man now at Coomb, 60 Cows for £3 a cow. 4 Horses carried Gravel.

5 March [Sunday]. I Went to Stoke Church in forenoon. Text 1 Corinthians 4.18 and they Sung Psalms 66, 13, 103. After Dinner Went to Wool Church to singing. A Pleasent day.

6 March. Of Thomas Willshear which he Owed me on a Bond £20 0. 0. I Rode to Pidle Early. John Allen and his Boy rited the Mill Wheel. After Diner I rode to Grange and saw some Cows but did not buy any. 4 horses Plowed in Long Craft and 4 carried Gravel to Rack Mead Wares. Wind at North with Rain and very Cold. Father sold a Horse to Mr Mermouth of Swannage for 22 Guineas.

7 March. I drew a Plan of Bovington Wheat Barn for Mr Frampton. Rode to Pidle called at Bovington as I Went. Met Father in Wooll Common who told me Mr Cockram Was there yesterday and that my Niece, M. Saunders, who lives [as] a servant to his Wife [Mrs Cockram] turns out bad. 4 horses Plowed in Long Craft and 4 carried Gravel etc. I Went with John Allen to Beer and laid there and sent home Daniel with some Meal.

8 March. I came with J[ohn] A[llen] early from Beer to Piddle. There we carried Gravel out of Furzegoalds Main to Rack Mead Weares with 2 Carts and 4 horses Ground at the Mill all day. Cold but not so cold as yesterday. I came home about seven. Brother H[enry] Loxly was yesterday at Blandford fair and came last Night to Bovington and hither this forenoon and departed about 2 so that I did not see him.

9 March. 4 Horses ploughed and 4 Made [an] End of Carrying Gravel Out of the Main. I rode round the Fields and Measured Snooks Lot which they routed in Hazzle Wood and which is 1 acre 3 roods 38 perches. T. Bascomb Accounted with J[ohn] Elliot and ows him £4. 7s. 10d. To carry 10 Bushels of Pease at 6s. [a bushel] to Piddle for Farmer John Homer at Beer. To be carried tomorrow. From out of Wood[street] I rode to Bovington and saw my Mother and from thence to Piddle and gave Orders to the folk. Coming home staid at John Elliott's and Paid him 21s. 6d. for Malt.

10 March. 3 Horses carried Gravel to Farm Mead Gates and 5 brought home Wheaten Meal about 6 sacks and bedded up the Barton. And about 3 Set out again with 20 bushels Peas and 12 Bushels and ½ Barley, Some Bread, Drink, Plants and Dust. I, D. Cobb and J. Lambert Winnowed Barley. Fine Dry Wether. Mr Pike came to see Pease But I had not any more to spare. Father called here but did not come in, Father seemed not pleased because I had the Mill rited.

11 March. 4 horses with 2 Carts Made an End Earthing 5 Acres and 4 Ploughed in Long Craft. A hard frost last Night and Snowed all the forenoon. Afternoon Dry. I rode to Wareham Market saw Farmer Willshear of Grange who paid me Use 0. 19s. 0d. And he Agreed to bring me next Tuesday 3 Cows and Calves for £14. 16s. 3d. and another for me to see and Buy if I Will. Father and I Dined with Mr Giles Brown and he shewed us an Experiment on Malt. See my shorthand Journal.

12 March [Sunday]. Sent Daniel [Cobb] to Wareham after my Pocket Book which I left yesterday at Mr Chisman's. I and Nanny went to Wool Church in Afternoon. Text Corinthians 4.18. We sang Psalms 66, 43, 12/33. W. Barnes asked for help at Church, agreed to give him some shoes.* We staid after Prayer and sung with our Tutor. Mild Weather.

13 March. 5 Horses carried a Load of Frith from Bovington to Piddle and 3 made End Plowing Long Craft. I rode to Piddle [and] Paid my Work folk there and agreed with Joseph Boit for a Reek of Hay at 27s. 6d. a Tun. Just as I got home Father came and staid a little. Nanny rode to Heffleton in the Evening. I rode thither and came home with her, got home between 9 and 10. A dry day but cold air.

* James Warne is here acting in his capacity as an Overseer of the Poor for Wool parish.

14 March. To Farmer Thomas Willshear for 3 Cows and Calves £14 16s. od. Of Farmer Homer of Shitterton for 10 Bushels of Pease £3 os. od. 6 Horses carried a Load of Frith from Bovington to Piddle and 2 rested. It rained in the Morning and most of the forenoon [*sic*]. I rode to Piddle in the Morning. Farmer Willshear to Bring Cows and Calves. He brought 'em and Met John Galton there, we talked of Changing Mares. My Colt, which I had at Woodbury hill fair and which is with foal, for a Mare which Father bought about 12 Months ago of Mr Lester of Poole and let Galton have last year. She's about 10 year Old and not likely to Breed. I would have changed for 16s. but he bid Me but 10s. 6d. so We parted. When I got back here was Farmer T. Willshear and his Wife who brought the Cows and Brother Florence and Sister. The 2 first departed soon after dinner and Brother Florence and Sister about 6. Cold Weather.

15 March. 5 Horses carried a Load of Frith from Bovington to Piddle. 3 Horses made an end restering 5 Acres and harowed some of it. I rode and carried some Victuals to Bovington for the Wagon to carry to Piddle. I staid a While with my Mother. Father set out this Morning for Ringwood Market. P. B[urden] made Cow Cribbs. Mild Wether. Chid Old Daniel Cobb for giving [up] of work so Soon.

16 March. 2 Horses rested and 6 carried a Load of Frith from B[ovington] to Piddle. I rode to Piddle [and] carried some [haw] Thorn Plants and let a Tut Jobb to T. Hunt and William Curtis to throw 2 Ditches, 3 feet wide and to feet deep Each, the Bank to be 5 feet [and a] ½ Wide at Bottom, to plant it and make a Hedge at Each side for 10s. a lug. I to find plants. This is from the Garden to the Lane going to Beer to make it straite with the Hedge between the Orchard and Conigar. Also let the Hedging of Lillington's Mead from Alner's Mead to Fords Garden, and from Rack Mead Weir to Brockhole Mead to Charles Hodges at 5d. a Lug. To Throw a ditch in Inside 4 feet Wide and 2 deep. Som[e] Loud Thunder and pleasent Mild rain in the Evening.

17 March. 3 Horses plowed in Mountfield and 5 carried another Load of Frith to Piddle. I rode to Bovington and spoke with W. Locass. At Woolbridge Met Father coming hither and here met Farmer Thomas Hayne of Coomb and Edward Masters Dairyman at Bindon and I Drew Articles [of agreement] between Father and them and they signed the same to Each Other. They departed between 7 and 8. Father came home from Ringwood last night. Foggy Mild, pleasent Wether. Wednesday Mr Frampton was at Piddle with Mr Cartwright and saw Piddle house. The same day I bled a Calf to death. John Davis of Blackpeston fetched Peas.

18 March. 6 Horses carried a Load of Frith from Bovington to Piddle and Harowed fallows in 6 Acres by Hodges. I rode to Wareham Market. Wheat at 6s. 6d. a Bushel, Barly fell, did not care to give above 25s. [a] Quarter. Ray Grass was sold for 10s. and 9s. a Quarter. My Cowgrass which is between 4 and 500 is all bespoke at 7d. a lb. It rained from Morning till After I got home from Market. A Pleasent Moderate rain, Blessed Mild. Read [?Received] a Letter To Father From Uncle R[obert] Duel's Wife.

19 March [Sunday]. I Went to Coomb Church in Forenoon as did more of Wool singers, and some of West Lulworth, who was Expected. We sung before they came Psalms 108 and when They came Joyned Company and Sung Psalms 9[a], 26[a], 100, 122[a], 16[a]. Father and J[oseph] K[ingston] Was at Church and R. L[ocass] who came from Piddle in the Morning brought round Father's Horses and He came with me and Dined here and departed about 3. Text at Coomb Philippians 3.15. Blessed Mild Wether.

20 March. 4 Horses Restered in Mountfield and 4 Carried Earth at Buckshill down into the Flatt. I called at Bovington as I rode to Piddle and carried my son, J[oseph] K[ingston] W[arne], with me. Father talks of going up to Totton next Wednesday to see about Uncle Robert Duel's affair. I saw Mr Frampton on the New road who liked the Plan of the Barn I drew for him and would have me get Farmer Jarvis to let the Old Kitchin be pulled down before Holliry day that they Might begin about rebuilding it the sooner. I called again at Bovington as I came back and borrowed Father's Mare and potts to carry Trees, Herbs and Flowers to Piddle tomorrow.

21 March. It rained from Morning to Night [so] that I and Nanny did not go to Piddle, but it not raining so fast after Dinner I rode to Farmer Willshear's at Grange to see some Cows and Calves but I did not buy any for they was Poor and not very Tite. I had a very Wet and cold ride, got home between 6 and 7. Snowed soon After and the Wind very high. Horsed [sic] did nothing this day.

22 March. It rained again in the Morning. Rivers very high with Snow which Melted. William Locass came home about 9 in Morning. Robin sick at Piddle. Sent William again with Flowers, Herbs and Currant Trees. I rode After. Met Father in Wool Common and called and saw Mother at Bovington and rode to Piddle and gave Orders what to have done and let a job of Banking, about 6 Lug, to Joseph Boit. Poor rates at Wool as I came home. Foggy Weather but Mild. Horses not Out. My little Mare Whitefoot 3 years old folded a Mare Colt this Night.

23 March. *[In margin the initials TH, GH, WC followed by 5 short-hand characters]* 4 Horses carried Earth at Buckshill and 3 restered in Mountfield. I and Nanny rode to Heffelton and from thence with Sister Patty to Piddle and set our Flowers and Herbs and then rode to Beer, put up at Cousin Fudge's and Bought things at Mrs Pearce's. Got home about 9. Mr Bestland was here in the Morning and have sold some Mow Burnt Barley for me to Poole for 24s. [a] Quarter and our Fat Pig for 6s. a score and appointed to send his wagon for a Load tomorrow. Mild Weather.

24 March. Rise Early. I, P. B[urden] and J. L[ambert] winnowed Barly, 13 Quarters 3 Bushels [and a] ½. Mr Bestland's Wagon carried to Wareham 8 Quarters of it and 249 lbs of Cowgrass Seed and is to come again for another Load tomorrow and the Pig which Thomas Blandford have Killed this Evening. 5 Horses carried a Load of Timber Gate and Posts from Bovington to Piddle and 2 carried Earth and Clots for the New Bank in Home plot. Father just called between 1 and 2 but did not stay. A little hoar [frost] in the Morning, but a Fine day. Farmer Talbot's Wagon called going from Grang and carried some dust: 2 Bags of Barly, 2 of Oats etc. 2 Piddle for me. William Locasses year out.

25 March. 7 Horses restered with 2 Sulls in Mountfield. I rode to Piddle in the Morning and After I had seen the Folk, Went from thence to Wareham Market. A fine pleasant day. I got home about 7.

Robin being better I sent him to Piddle in the Afternoon. Daniel came home He being about to Marry Soon with Elizabeth Snook. Stays with me by the Week, till George Sennick can come April 3rd. The last Night William Locass laid here at Woodstreet was between February 9 and 10th, 43 Nights since.

26 March [Easter Sunday].   Easter. I rode to Bovington and from thence Went to Moreton Church and dined with Mr Frampton. Text Philippians 3.10 first part. They sung Psalms 23, 92. Mr F[rampton] gave me a Plan of the House at Piddle and Told me to consider about it if any Alterations was proper to be made. He also told me They was going to fetch away the Mill Stones at Piddle tomorrow. I got home before Sunset. Nanny was at Wool Church. Text 1 Corinthians 15. 19, 20. They sung Psalms 103, 9ª, 48ª. Fine mild pleasant Weather.

27 March.   4 Horses Made an End Earthing of Buckshill and drugged an Ash in Mountfield out of the Way and 3 Horses Restered. I sent Daniel early to Piddle and ground some Barly I had there and Rode over thither and just saw the folk and about 10 Mr Frampton's wagon came and they took apart the Mill and carried away the Stones. I got home about 1 and Father came and We went to Church reckoning and continued William Snell, Churchwarden and choose my Father and Henry Brown in behalf his Father, Overseers. Blessed Mild, pleasant weather. From Church We adjourned to Clarks Ale. Emanuel Elliott wrangled with John Hayte, William Snell and R. Standly. I got home about 8 from Clark's Ale.

28 March.   Borrowed Mr Ekins's Wagon and carried 2 Loads of Frith Bought of C. Hodges from Stockly to Piddle. Yesterday Morning I Began to take off Mr Cartwright's Plan, but did not finish it, But finished it this Morning and began one of my own projecting. I rode to Pidletown Fair and called at Bovington as I Went but Father was gone. I carried some Cowgrass Seed to Mr Alner's for his Br[other] Bridge and dined there with them, Old Mr A[lner], Father White etc. Father, Father White and Farmer Talbot came home with me to Turners Piddle and looked on the House. We met with Mr Frampton and he rode back with us and says he'll consult Mr Cartright again on his plan and he Told me that William Talbot of Shitterton had been with him in the Morning and Offered to sell him Alner's living and he Bid me to write to Mr Derrick at Newport in the Isle of Wight about it.

29 March. 6 Horses Fetched 3 Load of Frith more from Stockly to Piddle. I rode thither and called at Bovington as I came back. Mr Frampton was there juste before me and gone again in his way to Farley [Farleigh Hungerford, Somerset]. In the Evening went up to Coomb Vestry but was to late. I was in at Farmer Barns's and took out an Account of Gates which they maintain (though I think wrong) out of the Church rate. I got home soon After 7. Blessed mild and pleasent Weather. My Father set out about 4 this Morning for Totton.

30 March. 7 Horses, 2 Suls made and End Restering in Mountfield. I rode to Heffelton and carried Father White a Clover Sieve. Mother W[hite] was coming up hither but Mrs Batrick of Beer came in and hindered her. From Heffelton I rode to Bovington, just staid a little with Mother. After I came home I and Nanny Went to Stoke intending to have spent an Evening at Mr Robinson's but He, his Wife and Brother Thomas was gone a visiting to Heffelton, so We went and spent the Evening at Rushton with Farmer Charles Burden and his Wife till 10 Clock. Blessed Fine Weather.

31 March. Horses ploughed with both Suls in 5 Acres at Buckshill for Pease. I rode to Bindon Abby and Saw George Northover, James Bascomb and his 2 sons drawing out of Stone out of the Old Ruines to build a Bridge and Foot the Cow pen Walls with. From thence I rode to Piddle and saw my work folk and to Beer and Bought some things. I staid a little at Cousin Fry's and came back to Piddle Church reckoning, where they chose me Overseer [of the poor] with George Pain. I called at Bovington as I came back, Father was not returned. Fair Wether continues.

1 April. Horses plowed and dragged peas in 5 Acres. I rode to Wareham Market. Met my Father there who came this Morning From Uncle James Duell's at West Moors. He went to Totton with my Father and Uncle. I rode to Winton and saw his Brother, Uncle Robert Duel. Father says they ows much more besides what he is in Gaol for. Father gave Aunt, Uncle Robert's wife, 4 Bushels of malt and ½ Hogshead of Beer to keep on and help the Family while He is in prison. Thomas Pike of Cold Harbor paid me the rest of Mr Brixy's Mony, £23 and 25s. the Use £26. 5s. Blessed Fine Wether continues. Corn seems to Sink.

2 April [Sunday]. I and Nanny rode to Wareham Meeting. The 1st Time she've been there since the Small pox. Text 2 Thessalonians 1.12 *[shorthand characters].* B[lessed] Sacrment was Administered in the Afternoon. This Afternoon young Farmer Smith of Ho[l]me was buried at Stoke. Father White and Brother Stephen was at Wareham. We dined at Brother Florence's. My son J[oseph] K[ingston] came from Coomb Church and stayed here. Pleasent Weather, but Cloudy.

3 April. 3 Horses ploughed at Piddle and 4 came home with the Wagon, brought some Oats 3 Quarters and 4 Bushels of Raygrass for Farmer Talbot. Bedded up the Barton and fetched in Furze and laded up dust and 3 Quarters Oats and 4 Bushels of Barly and after dinner set out again. George Sinnick of Woolbridge to come this day in William Locass's Stead. He came and went to Piddle with the Wagon. I, P. B[urden], J. L[ambert] and D. C[obb] Winnowed Barly. I and Nanny went to Stoke and spent the Evening with Mr Robinson and with Father and Mother White, Cousin Elizabeth Fry and Sister Patty, which last 2 came hither with us. Light in the North at Midnight.

4 April. I rode to Piddle in the Morning. 6 Horses fetched a Load of Frith From Bovington. I called there as I came back and Father came with Me. Farmer Garland cut Father's Lambs. Mr Pike and Mr Cockram came and We Went to Bindon and Mr Cockram and Old Farmer George Stickland Valued 23 Cows and 7 yearlings, the Cows at £4 15s. each and yearlings at 2 guineas Each for 6 and 1 for 1 Guinea only about 8s. on the Whole so that the sum is £122 10s. od. After they had seen the Cattle and viewed some Hedges Father, Mr Cockram, Farmer Garland, Mr Pike, Farmer George and Farmer Philip Stickland all came and dined here and Staid till about 6. Sister Patty and cousin Fry Departed about Sunset. Cold Air and Showers of Snow.

5 April. 4 Horses carried Earth at Piddle and this day began sowing Oats in Longcraft. My Dairyman Philip Whittle came and saw his dairy Cows and We signed our Agreements. Father was just in here about Noon. I Went round the fields and after to Bindon and saw the Dairyman Mark the Cows which Father are to have and I rode to Wool paid for Some Malt etc. Several Cold Showers of Snow and Hail but a Very drying Wind. Paid Daniel Cob his year's wages and he now goes on a 4th year.

6 April. I rode to Piddle. 3 Horses Ploughed in 5 Acres and Harrowed long Craft and the Pease. I Measured the Banking which T. Hunt and William Curtis have done between Conigar and Home plot. I agreed with Farmer Jarvis to let the Masons pull down the Old Kitchin which they'll begin about next Week, and I do Acquit the Farmer from doing anything about the Water work in Gully. I called at Bovington as I came back, staid there but a little. Such Wether as yesterday. I called at Heffelton as I went, carried Mother white Some Lambstones. Mother W[hite] came in the Afternoon.

7 April. Of Farmer Wood and Mr Davis for keeping 168 sheep of which 5 died at Piddle £25 0s. 0d. I rode to Piddle early and Met Farmer W[ood] and Mr D[avis] there and they had away their sheep. I saw the workfolk and gave Orders and brought home sacks with Me. I measured another Lot in the Wood routed by the Snook's which is 1 Acre *[space]*. A White Hoar in the Morning and Fair and Dry day. 1 Horse rested and 6 Fetched a Load of Frith from Bovington. Went to Church to Singing in the Evening. Nanny went to Bindon to See Mrs Brown but she was not at Home.

8 April. I Measured up Oats in the Morning and After rode to Wareham Market and carried the Butter. 3 Horses plowed at Piddle and 4 carried a Load of Bushes from Stockley to Piddle. William Locass came home in the Evening and I paid him for a year and 2 Weeks £6 6s. 0d. For William Locass place of Residence or Proper parish, See March 25. I think He belongs to Turnerspuddle For He hasn't laid here at Woodstreet Since February 10th till this Night.

9th April [Sunday].   Prayer at Wooll in Forenoon. I, Nanny and Son, Stephen, Went to Church. Text Samuel 3.39. We sung Psalms 108, 21, 26ª. The B[lessed] Sacrament at Wooll which my Father, Mother and I with Many more stayed and received. After Dinner I went to Church again. Met Some Singers and Elias Hibbs our Tutor and sung together for some Time and talk of Going to East Lulworth Church next Sunday and to Moreton this day 3 Weeks. Joseph Davison and his Wife here in the Afternoon.

10 April.   Sister Brown's 3rd son Born at Martin. Farmer Talbot's Wagon Stopped here going from Grange to Piddle and carried along some Dust, 28 Bushels of Oats and 20 Bushels of Barly for me, 3 Stocks of Bees and some Victuals. 4 Of my Horses carried Timber from Bovington and 3 ploughed at Buckshill. William Locass do Stay a Week or 2 More with me so He set out in the Morning to Piddle and there bides. My Father called here about 11 but did not stay long. Afterward I rode to Piddle and saw the Folk and Paid the Labourers. Fine pleasent Weather. John Allen put up the New Bars by Robert Philips's.

11 April.   I Met George Sexey, George Read and John Hood etc and my Father met us there and We measured all the Lotts that was not measured before and I make it to be in all about *[space]* A. *[space]* R. *[space]* P., the routing of which at 3d. a lug coms to *[space]* and *[space]* lug Levelled by G. Read at 4d. makes the whole cost to be *[space].* About 7 or 8 in the Morning Wind at South, it began to rain very pleasant which soon Abated but continued till near Night. Horses ploughed and Sowed Some Oats at Buckshill. After Dinner I rode to Piddle and called at Heffelton and took Measure of their Pickling Tub, and sent to Cousin Fudge of Beer to make me One. I drew another Plan for Piddle House.

12 April.   I rode to Piddle Early. The 7 Horses made an End plowing and sowing of Oats and Pease at Buckshill and finished dressing the Ground. It began to rain again a little. About 1 I Ordered the Wagon to be at Home tomorrow and called at Bovington as I came back and Accounted with my Father and received £9 9s. 7½d. of him. Farmer Jarvis (though Unwillingly) began to Clear out their things out of the Old Kitchin and J[ohn]

Allen began to take up the Loft over it and the Masons intend to uncover it tomorrow. I Weighed Hay with Joseph Boit and William Curtis the last I shall have of him which is 4 Tun and ¼ at 27s. 6d. which comes to £5 16s. 10½d.

13 April. I, D. C[obb] and J. L[ambert] Winnowed about 10 Quarters Barly. And Afterward took in cut of Ray grass. George Sinnick and Robin came home with 6 Horses and the Wagon and carried Dust, 7 Quarters Barly, 12 Bushels Oats and other things to Piddle. A brisk Wind at East and very drying.

14 April. 3 Horses turned back fallows in Pit Close and 4 with 2 sulls plowed in Lower Hoglease and We began Barley Sowing, Sowed 8 Bushels. I rode thither and Saw how they went on and from thence to Winfrith sessions and had the Poor Book confirmed and got home about 5. Went to Wool in the Evening to Meet the Singers but was sent for Home a Cow being Ill in the rising. Dr Barratt came up, broke the Bladder under her Tongue, raked her and let her blood and the Cow done Very Well again.

15 April. The 7 Horses Plowed and Sowed as yesterday. I rode to Blandford Market and Sold 10 Quarters of Raygrass seed there at 10s. [a] Quarter. Clover was Sold at 4d and 4½d, all Wheat 6s. 6d. and 6s. 7d. a Bushel. I saw and Dined with Brother Roger White and Brother Brown of Martin who told us Sister [Brown] was delivered of another son last Monday Morning. I called on Mr Cartwright at Blandford and Shewed him my Plan of Piddle House which he approved of and Would Send to Mr Frampton. At his house I saw Esquire Plydell of Milborn and Asked his leave to dig some Earth in Shitterton Heath, which he granted me.

16 April [Sunday]. As We came from Blandford Market last Night it Snowed a smart shower which laid Thick on the Ground this Morning, with a hard Frost But the Son shone Fine and it soon went off. I and Nanny went to East Lulworth Church and Met Wool Singers There. Text John 16.22. We Sung Psalms 21, 16[a], 96[a], 132[a], 26[a] and the Communion Responsals. I and Nanny Dined with Farmer Garland and his Wife and Mr Lee, a School Master now coming to set up at East Lulworth. We got home about 5.

17 April. Horses ploughed with 3 Suls and Sowed 12 Bushels Barly in Hog Lease. I and Nanny rode to Wareham Fair. We Saw Father and Father White there. We was at Brother Florence's and Dined and Drank Tea. I Bought a Book, *The Art of English Poetry** and a Large Bible which cost 12s. 3d. the Other book 12½d. Cattle sold dear. A Cloudy Morning. Rain all the Middle of the Day and a Fine Dry Evening to come home in. We got home about 8. Saw Mr Harvey of Chichester at Wareham Fair.

18 April. The 7 Horses Ploughed with 3 Suls and Sowed 12 Bushels more of Barly. I rode to Piddle and Measured Hedges which Charles Hodges have made and paid him for it. Nanny rode to Heffelton to see her Mother and meet Sister Florence there and as I came back from Piddle I called there and We came home together about 8 at Night. A dry Forenoon but about Noon the rain came on and rained till night.

19 April. G[eorge] Sinnick, William Locass and Robin came home with the Wagon and 5 Horses (the other 2 rested) and carried a Load 10 Quarters of Ray grass Seed to Farmer Brine's at Chettle down near Blandford, they Arrived between 2 and 3 and set out about 4. W[illiam] Locass went and gardened for himself at Wool. I went to Bovington after dinner and fetched an Old Horse of Brother Michael Saunders's and had it Shoed, and intend to drive him a little this season. G[eorge] Northover, James Bascomb and his son James and my Father with them dug up about of 18 Inches deep between to of the North Pillars of the Church a large Tomb stone long – 7 feet 5 inches, Wide – 4 feet, Thick – 6 inches. My Father intend to beg it of Mr Weld.

20 April. W[illiam] Locass set out early to Piddle with Brother Michael's Old Horse and ploughed and sowed 4 Bushels of Barly with 2 [horses] and G[eorge] Sinnick and R[obin] Locass came home with the Wagon and 5 Horses and carried 5 Sacks of Barley, 2 or 3 of Oats and 10 of Raygrass seed and Father's Old Roller of Bovington which was up here. And when the Wagon departed I rode to Piddle and saw how things was there, I got home about 8.

* An anthology compiled by Edward Bysshe which ran to many editions, the 7th published in 1724.

Farmer Jarves paid me £9 2s. od. the Award Money towards making up the Hedges there. Mr Robinson, his Wife, and Brother Thomas and Farmer Charles Burden and his Wife came to see us about 5 o'Clock and Staid till Midnight. A Blessed Fine day, Mild and Warm.

21 April. 4 Horses dragged down Resters in Mountfield and 3 made an End Plowing 2 Ridges that was begun, and sowed 4 Bushels Barly and 36 of Grass Seed and Cross Harrowed it in which is ½ the Lower Hoglease. John Allen put the new Bridges by Gully. I called at Heffelton as I went to Piddle and had 36 lbs of Clover seed at 4d. and I called at Bovington as I came back and saw my Father and Mother. Father intended to go to Blandford Market tomorrow to Meet Uncle James Duell there. Dr James came to Bovington just before 7 and I came away soon after. Mild Weather some few drops of rain. The Masons have got the Old Kitchin at Piddle so low down that J[ohn] Allen helped them down with the Chimny piece yesterday.

22 April. 4 Horses Dragged as yesterday, and 4 Ploughed with 2 Sulls and sowed 8 Bushels Barly. I rode to Wareham Market with the Butter. I Ventered my Chance for a Packet of Dr Axford and a Silver punch Ladle he Gave which was got by a man of Wareham. I got home between 8 and 9. Fine mild pleasent Weather. Wheat Sold for 6s. [a] Bushel and some 2d. more. I sold 10 Pints Kidney Beans for 3d. a pint to James Brown and left with him for He to sell for me 5 pints Bunch Peas and 2 pints of Black Eyed at 2d. a pint.

23 April [Sunday]. I and Nanny went to Wool Church in Afternoon. Text Psalms 41.4 and we sung Psalms 128[a], 96[a], 117[a]. My Father and son J[oseph] K[ingston] was at Church, J.K. in his new Cloths. Robert Standley's Wife and Robert Stickland's Wife of Burton came with Nanny from Church and R[obert] Standley came with me and about 5 R[obert] Stickland came and they all stayed till about 8 Clock. A Very brisk, drying Wind at East. Would [have] been very hot had there been no Wind. I received a Letter from Mr Kingston for my Father to meet him at Lulworth Castle next Thursday Morning.

24 April.   W[illiam] Locass came home Saturday night and went to Piddle again this Morning. I rode After and had along the Young Mare and Colt which I intend to drive now and then a day. 4 Horses ploughed with 2 Suls and sowed 12 Bushels Barly, 4 made an End draging in Mountfield. Made an End Threishing and Winowing Barly which was much better this last year than the 2 first years as might be seen under in

   1755 but little more than 10 Bushels and a Peck in an Acre
   1756 a little more than 13 Bushels 1 Peck and 3 Quarts in Ditto
   1757 about 21 Bushels 3 Gallons and $\frac{1}{5}$ of a Gallon Ditto.

I Agreed with Robert Philips for he to rout the Furze in the little Plott belonging to his Bargain for 2s. and 6d.

25 April.   I called at Bovington and saw my Father and Mother as I rode to Piddle. 4 Horses Ploughed with 2 Suls and sowed 8 Bushels Barly and 4 made an End Harowing in Hoglease. I brought Bags to Heffelton which I borrowed of Father White and there was Mr Robinson, his Wife and Brother. I staid with them till 8 Clock. Father brought up Mother and left her here. Drying Wind. The 2 Sulls made an End plowing Hoglease and We stroke out 22 Ridges at the North of Mount Field. Finished Sowing Clover in Hoglease.

26 April.   I rode to Piddle. 3 Horses Ploughed and sowed about 4 Bushels B[arley] in M[ount] Field. 4 Horses Finished Cross harrowing Hoglease and Cross Harrowed Some in Mountfield. Father made an end sowing yesterday at Bovington and began here this day in Knap and Wood Close. Father call[ed] here before I went to Piddle and had not been gone long 'ere Father sent for Mother for he heard at Wooll that Mr Robinson, his Wife and Brother Thomas with Sister Patty White was all gone to Bovington, so Mother went away before Dinner. I got home between 7 and 8. Wind at North and Cold Small rain in the Evening.

27 April.   *[space]* Horses ploughed with *[space]* Sul in the Mount of Mountfield and Harowed After. R. Squire with his cart carried 16 Bushels B[arley] and 32 of Raygrass Seed to Piddle for me and was to have brought back but was late and so did not. Between 8

and 9 Father called on me and I rode with him to Lulworth Castle and there with Mr Weld, Mr Kingston and Farmer Garland settled the Leases for Woodstreet and Bindon and I paid Mr Weld my last Ladyday's rent. When We got back My Father had his little Mare and let me have Young Jugg and I am to allow him 8 Guineas for her. Not much rain last Night. Wind at North, cloudy and Cold. Farmer Talbot at Piddle Married to his Servant Maid, Niece and God Daughter, Elizabeth Toms.

28 April. The Wagon came Home with 5 Horses and carried Dust, 4 Bushels Oats and 8 Bushels Barly for the Horses and *[space]* Quarters to Sow, a Bed etc. and some Straw. 3 Horses Ploughed and sowed 4 Bushels. I rode over and just saw how things was and called at Bovington as I came back and there was Uncle James Duel just arrived. And I came to Wool and there at the Vestry delivered up the Poor Book to Henry Brown and After we broke up He, I and William Snell went and Took an Account of H. James's Goods but He had Sold it before to his Daughter for 10s. She shewed us a paper thereoff and I took a Copy of it. Brother Roger and Brother Stephen White came and spent the Evening till near midnight. I and Nanny went to Coomb last night and bought 3½ Hogsheads of R. Meech for 12s. I received a Letter from William Derrick at Newport about William Alner's living at Turnerspiddle.
[inserted in red ink] J.W. Born April 17 Old Style 1726.

29 April. I rode to Bovington and from B[ovington] to Blandford with Uncle James Duel and there Bought a Cow (and Calf) 4 years Old for £3 19s. 6d. and Mr Nipprod's Men of Abbs Court brought it thither for me. I went with Uncle J[ames] Duel to Mr Ridout's and drew a contract between them for Pottern Farm which Uncle J. Duel is to enter on at Michaelmas next. I there wrote a Letter to Brother M[ichael] Saunders at Martin to provide me £40 he owes me by the 9th of July and another Letter to Mr Frampton at Farley Castle with a Copy of Derrick's Letter to Me. 3 Horses Rolled all the Peas and Oats on the South Side of the Water all which W. Locass sowed with Turnips in the Morning, and 3 horses ploughed and 2 Harrowed in Mountfield and sowed 4 Bushels Barly. I came from Blandford about 7 with Mr Nipprod to Abbscourt and drank a dram with him, got home about 10. Cold Air.

30 April [Sunday]. I and Nanny and S[tephen] W[hite] went to Stoke Church in Forenoon. Text Revelations 3.19 and they Sung Psalms 18, 116 and 139. Father was there. We was just in at Mr Robinson's but not to stay. After Dinner I Went to Coomb Church, Text the Same, no Singing there. I and Mr Robinson Stayed at Mr Bewnel's till 5 Clock. Mild and Cloudy.

1 May. 6 Horses Ploughed with 2 Suls and 3 Rolled and harowed and Sowed 10 Bushels of Barley. Farmer Talbot's Wagon came and carried some Dust and 4 Quarters 2 Bushels Barly and 4 Bushels Oats for me to Piddle. I and Daniel Cob drove 13 Heifers to Piddle to Grass and to have 'em kept out in the Heath at Day and put into Ground at night, 'twas 7 yearlings 3 2 yeared [old] 3 3 year Old Each, 1 of the 2 yeared ones is with Calf. I called at Bovington as I came back and Father and Mother would have I and Nanny to be there to Dinner tomorrow with them. I bespoke some Hay of Farmer Robberts at Droop to give him 27s. 6d. a Tun for it. Mild, Warm Forenoon, Colder afterward.

2 May. [space] Horses ploughed with 2 Sulls and *[space]* Harrowed After W[illiam] Locass sowed *[space]* Bushels Barly and some Turnip Seed at the North of Mountfield. Father came up round here and just called in. Sister Patty White came about 10 and between 12 and 1 She, I and Nanny Rode to White Way and Dined and drank Tea at Mr Alner's. Mr Alner, Mrs [Alner] and their Daughters Kitty and Nanny, and 2 Grand Daughters, Molly and Jenny Clavells, where *[sic]* all at Home. We came from thence about 6 and called at Lulworth Castle and I asked Mr Weld's leave to carry away what Hay I shall have left After Holliry day and he granted it. My Wife and sister went in too and Betty Vin shewed us the Castle and Mrs Low Treated us very Handsom. Fine pleasent Weather. We got home about 9.

3 May. I rode to Bovington and called on my Father and he rode with me to Piddle. 5 Horses ploughed with 2 Suls and 3 Harowed. Farmer Alner sent a Sull and 3 Horses yesterday about 11 and all day to Day. Finished the North Furlong and sowed *[space]* ridges in the South F[urlong]. Father and I staid in a little

44

at Farmer Alner's and From thence to Affpuddle and Father received £20 of Thomas Saunders and 2 years Interest, due on a Bond which Father gave up. While we was ab[sen]t Nanny called on young Mrs Standley and went and dined at Bovington with my Mother and was just gone to R. Stickland's at Burton. I bespoke a Load of [space]. My Father and [I] rode from Bovington to see if Father White had done sowing. He Finished this day and his wagons to Carry my Load of Barly from Bovington tomorrow. Fine Weather. From Heffelton I rode to Burton and at R. Stickland's met R. Standly and our Wives and came from thence about 11, got home about 12 and here was Uncle John White, Their Maid and Dairyman's daughter, they departed about 1.

4 May. Father White's wagon carried 2 Sulls and 9 Quarters of Barley from Bovington and in Bovington Heath broke the pillar of the Wagon, so went to Piddle and got my Wagon and carried it. Farmer Alner helped us again with 2 Suls and Finished plowing Mountfield into 2 or 3 ridges. We ploughed with 2 Suls and [space] Horses Harowed after Sowed 24 Bushels. Farmer Mate of Beer sent me word He would send some Horses to Help me. I called at Bovington as I came home. Fine weather Continues.

5 May. Farmer Mate sent 4 Horses and 2 Suls and Father White's 6 Horses and 2 Suls and with our Own 2 Suls and 4 of the Horses 4 being cross harowing in Mountfield the[y] finished Plowing and sowing of Mountfield which Took about 10 Bushels More and We ploughed a good piece in Spear's plot and sowed 16 Bushels there. Cold in the Morning and some rain about Middle of the day but not a great deal, a mild Evening.

6 May. I rode and carried the Butter to Bovington and so to Piddle. Farmer Mate and Father White sent their 10 Horses again which finish[ed] that Plot and Pit Close. Spear's Plot took to sow it about 24 Bushels and Pit Close about of 10 Bushels. A Blessed Fine Morning and day. I rode from Piddle to Blandford Market bought a Cow and calf there cost £5 2s. 6d. and 30 Chilver Hogs and 7 Couples which Cost £21 19s. 6d. From Blandford I rode to Wareham Market and tryed my chance for a Silver spoon but did not get it. I got home [between] 8 [and] 9 Clock. There was 3

45

spoons given one got by G. Grose of Wareham, one by Robert Stickland of Burton and one by John Barnes of Litchet.

7 May [Sunday]. My Man William Locass was married early this Morning to Elizabeth Snook at Wooll. About 9 Farmer Garland and his Wife of Lodge came to see us. After Dinner we went to Wooll Church. Text Revelations 3.19. We Sung Psalms 149, 48ª, 8/128ª. The Parish Granted John Hurst a Certificate. After Prayer my Father came and Elizabeth Squib. Father and Farmer Garland and his Wife departed between 6 and 7, about that Time Sarah Brown and Betty Allen of Wool came and Some Time after Richard Squib to go home with his Wife. They all departed between 8 and 9. I just went to William Locass's after I came out of Church. A little rain but not much.

8 May. [Borrowed from] Uncle T. Saunders of Affpuddle a sul and 3 Horses to help us which, with Farmer White's 2 and my 2 Suls, almost ploughed that Plot by Charles Hodges and W. L[ocass] Sowed the Great Plot West of the Water, almost 24 Bushels. Some made an End Cross harrowing Spear's and Pit Close. I called at Bovington as I returned from Piddle and staid till after 7. T. Barrat invited me and my Wife and My Father to his Wedding which is to be tomorrow. Some few drops [of rain] in the Morning but quite hot Afterward.

9 May. Last Night and this Morning W. L[ocass] Sowed Spear's and Hodges with Turnips and 4 Horses made an End dressing it. Sowed about *[space]* Bushels of Barly in that Ground and Father White sent his 2 Ploughs of 6 Horses again. Brother Stephen has been every day with them. Made an End ploughing and sowing the little Plot and Ploughed with 1 Sul of my own, Father White's 2 and Uncle T. Saunders' 1 *[space]* the Turniplot and sowed about 3 Bushels [and a] ½ of Barly and *[space]* Bushels of Oats in it and hope to make an End sowing and Finish it all off tomorrow. I Brought back sacks to Heffelton as I came back and stayed at Thomas Barratt's Wedding. He was Married this day to Elizabeth Knapton of East Lulworth. My Wife went down in the Evening and We broke up about 9 Clock. Nanny was frighted coming home at a great Dog that run at her in Pear Close and I saw a man

46

lie down there whether Drunk or not I can't say – it was George Sexey of Hywood, Drunk.

10 May. Father White's 2 Ploughs, 6 Horses helped us again but did not quite Finish, they had home their wagon and suls, 1 sul of my own ploughed all day. 3 Horses Ploughed, Rolled and Dragged and 2 Horses Harowed. Afternoon very Warm. The Masons at Piddle have got out the Foundation for the House and laid but not Settled the Corner Stone. They are to Build after Mr Cartwright's 2nd Plan but the Mason had rather build it by mine. But Mr Frampton not being at Home I came to Moreton with *[space]* Jarrit to speak with Mr Lockyer about it but he was not at Home.

11 May. 5 Horses Made an End Plowing and Sowing in the North close and 4 carried in some Turves and Furse for myself and some Furse for the Dairyman. I rode over to Piddle and called at Bovington as I Went and came back but did not See my Father for he was not within when I Went and was gone to Cliff with Farmer Edward Sabbin and his Wife in the Evening and not returned. Some rain about Noon but not much. Mr Lockyer met me at Piddle and Saw how the Workmen went on and said he would write to Mr Frampton to have down the Old Porch. Mr Bewnel and I came from Bovington together.

12 May. 6 Horses carried a Load of Frith from Bovington to Piddle 3 rested. I rode to Bovington, carried Old Iron to Wool. Father was round this Way and I missed him. He went to Coomb and dined with Mr Bewnel's Sisters. W[illiam] L[ocass] and R[obin] both came home in the Evening with 4 Horses. Joseph Davison and his Wife, Thomas Barratt and his Bride, Christopher Barratt, Henry and John Brown came to See Us. The 2 first Went away when the Other came, and they stayed till near 11 Clock.

13 May. Rise Early. The Folk with the 4 Horses and Waggon set out from home to Clifft and G[eorge] Sinnick with the Other Wagon met them there and carried my Dairyman's Goods from thence to Piddle. I sorted out the Cows and with Daniel [Cobb] and William Hunt Drove the 20 Dairy Cows, and a 2 year Old Bull with them to Piddle and from thence rode to Blandford Market, on the Rode met Brother M[ichael] Saunders, going to Bovington with his Daughter Elizabeth. I Bought no Cattle at Blandford. I was to late for the sheep, and the Cows as well as Sheep was very dear. I came from Blandford to Wareham Market with Mr Cockram. Met Father and Brother M[ichael] S[aunders] there and came home with them. I thought to have sold some Wheat and Barly at Wareham but did not.

14 May [Sunday]. I and Nanny rode to Bovington and carried S[tephen] W[hite] and J[ames] L[yne] with us. I, J[oseph] K[ingston] and S[tephen] W[hite] with several of Father's Men Went to Moreton Church. Text Acts 2.1, 2, 3, 4 and they sung Psalms 8 and 84. Father and Brother M[ichael] S[aunders] Went to Stoke Church. Text the same there and at Coomb in the Afternoon. Tom Lynes' Father of Burleigh and Farmer Effemay of Hinton Martin [?Hinton Martell] Dined at Bovington and Departed between 4 and 5. We Staid till near 6, left S[tephen] W[hite] there. Inclined to Rain in the Evening but not much. Cousin Richard Duell at Bovington was Married to his Fellow Servant, Elizabeth or Betty Coffin, this Morn at Wool Church as was Robert Mead to Jenny Masters.

15 May. I rode to Bovington and from B[ovington] to Piddle with Brother M[ichael] S[aunders] and was home again before 8 Clock. Had 2 Bulls and Heifer and Sold the Heifer and a 3 yeared Bull to be Baited at Winfrith next Wednesday for £4 3s. od., the other bull brought home again unsold. Father and Mother, Brother M[ichael] S[aunders] and his Daughter and Nanny, J[oseph] K[ingston] and S[tephen] W[hite], our Sons, was all there as was Father and Mother White, both Brothers and Sister Patty. 'Twas very fine Weather. A pretty large Fair* and Cattle in General sold very dear. Mr Cockram and his Wife was both at Fair and Mrs [Cockram] rode to Bovington and came down with my Mother, I and Nanny got home about 8. Ed[ith] Mead was resolved to go home, But We would not let her and she Howled prodigeously about it.

* Woolbridge Fair, normal date 14 May but in 1758, that being Sunday, it was held on 15 May.

48

16 May.   Yesterday 6 Horses carried a Load of Hay to Piddle and brought back Turves from Bovington Heath and Carried now Logs and Clift Wood and brought back another Load of Turves. I rode to Bovington in the M[orning] and Father came hither with me and soon after Farmer Philip Stickland and they Two rode together to Lulworth Castle. Father went to Pay Mr Weld for Bindon Dairy Cows (See April 4). Father intended to Call as he did come back, but sent word He should not. Brother M[ichael] S[aunders] talked of going home but Father persuaded him and he stayed and brought my Mother up hither and Elizabeth S[quib] came up. Made an End Threishing wheat and Winnowed it. I had this Year about 12 Bushels in an Acre. We took down some Beds this Evening and put things in Order to move off tomorrow. The Men are now catching up the Fouls between 9 and 10.

17 May.   *[written in the margin]* Came to Live at Turners Puddle.* Rose Early and before We had Laded both my Wagons 1 of Father's came to Woodstreet and Father came too before we was gone and helped. Jane Cob, Martha and Elizabeth Mead was there and helped too. I bought a Hog Colt of Joseph Davison for 49s. and brought it with us. We came from Woodstreet about 11, Drank at Wool with several as we came by. We got here between 2 and 3. Heavy Loads broke several Earthen[ware] things. Got in most of it but the house is in a grand litter. We brought 2 Hog Bulls, 12 Cows and Heifers and all the 17 [?7] Weanlings. Left all the pigs behind. Hot weather.

18 May.   Wagon fetched a Load of Lime from Norden by Corfe [Castle]. I Rode to Wareham and met it and got 11 Butter Tubs and Grease and sent home by it. I saw Father there, his Wagon carried a Load 8 Quarters of Barley to Mr Bestland for me. I called at Mr Manuel's going along and bespoke a Quarter of Oats which he sent me before night. C. Hodges, Daniel [Cobb] and William Hunt fetched all the Pigs from Woodstreet but was late 'ere they got home. They Also brought 3 Heifers and 3 Calves, 2 of the Calves I sold just After they Arrived to Farmer Randell for 20s. I had up one of the 20 Cows from Dairy house which I

* At Turners Puddle Farm, as tenant of Mr. James Frampton of Moreton House

brought from Woodstreet instead of which he kept 1 I bought at Blandford Market the 6th instant. Hot Weather Continues.

19 May.   Wagon with 5 Horses fetched 2 Load 1600 of Brix from Oakers Wood. I rode to Bovington and saw my Mother. Father was gone to View Fences at Creech between Mr Bond and Farmer Thomas Willshear. From Bovington I rode to Wool and Woodsteed *[sic]* and brought home a little wet Barley with me. I called at Heffelton as I returned and got home about Sunset. Hot Wether continues. I Ballanced Accounts going through Wool with Thomas Barratt and made Even. I carried Home Farmer Burden 2 Bushels Oats which Daniel borrowed there last Tuesday.

20 May.   3 Horses Rolled and made an End on the Livings. 5 Picked up Wood about and carried into Gully to make the Fence Good there. I caught 2 pretty Trouts down at Mill, Both about 3lb. I and John Allen Tited up and placed about some of the Beds and things and made it look much better than 'twas. I did not go to market any Where. Wether continued hot. About 2 this Morning Molly Hurl from Pottern by Verwood Arrived in Farmer Crisby's Wagon of Ower Who brought a Load of Earthen wear from thence.

21 May [Sunday].   [written in margin] Trinity.
   I Was going to Beer Church in Morning but there was no prayer so I returned. Between 10 and 11 Sarah Brown of Wool came and about 2 Mother and Brother Roger White, I and they 3, with Molly Hurl, Went to Church between 3 and 4. Text 1 John 5.7 and they sung Psalm 66. After Prayer the Poor was paid and a vestry held about Robert Philps. Mr Fisher is a comical Man to Preach and as Odd Way of Reading Prayers as ever I heard. He reads so fast. Mother White, Brother R[oger] W[hite] and Sarah Brown all departed between 6 and 7. Hot Wether Continues. R[obert] Duel bad, Nanny bid at home with him.

22 May.   3 Horses Rolled, 5 carried a Load of Turves to Woodstreet and brought home Dust from thence and thorns on Top from Bindon Abby to Robert Philp's Plot. I caught 2 More Trouts at Mill, scarce as big as those on Saturday. The Wether

Foggy in Morning but turned to heat, yet seemed inclined to Change for Thunder or Rain. J[ohn] Allen Rited up more things within. Hung up a Gate Between the Bartons and the Masons goes on Smart. Poor R[obert] Duel continues bad.

23 May. 3 Horses Rolled all Lower Hoglease and 5 fetched in 2 Load of Turves for the Dairyman and 1 for us. Saturday the Dairyman saw the Cow I bought last at Blandford piss Blood and it continued Bad so I rode to Beer and got some red Wine and Dragon's Blood and gave her and Since dinner carried her some toast and Ale. About Noon William Hurl (Molly Hurl's Father) came as He was Going home from Bovington. He laid at Bovington last Thursday night and has been since at Weymouth and Dorchester. He departed between 3 and 4. The Gamekeeper Brownjohn called in and 2 Women from Beer to bespeak Turves. No rain yet, watered trees and Flowers this Evening. Got my Journal up Even which was not before since I've been here.

24 May. 6 Horses fetched a Load of Frith from Bovington. Molly Hurl rode thither in the Wagon. We Gardened, Dug and Set Beans. While at Dinner, Mr Manuel and Powel of Forthington *[sic]* called but did not allight. After Dinner I rode to Bovington. Saw Father and Mother and Ballanced Accounts with my Father and brought home S[tephen] W[hite]. Got home abut sun set. Hot Wether Continues. The Child R[obert] Duel Bad and Also the Cow.

25 May. I rode to Affpuddle and Spoke to Farmer John Gould to come and See my Cow which He did in the Afternoon and gave her a Drench and We hope she'll recover. I called and Breakfasted at Uncle Thomas Saunders's as I came back had a little rain but not much. 6 Horses fetched in 2 Load of Turves and 100 fus fagots. Set more Kidney Beans. The Child R[obert] D[uel] we hope Something Better.

26 May. About 6 Mr Lockyer called here and soon after my Father, they staid a little While and I rode with Mr L[ockyer] to Woodburyhill and saw some Poles He bought there which I am to Fetch. And from thence I rode to Woodsteed *[sic]* Whether I had ordered the Wagon to. We brought home Some Dust, about 10 Bushels of Wheat, 4 Dozen of Hurdles, a Wash Tub etc. I staid at Wool and had my Horse shoed and dined with T[homas] Barratt and Henry Brown gave me some Apples and We Agreed to go to Moreton Church to Singing next Sunday. A Little pleasant rain.

27 May. 6 Horses fetched 39 pairs of Poles (which cost 2 Guineas) from Hove Coppice by Woodbury hill South Gate, and some more from Bryants Puddle. I Was with them at Copice and rode from thence to Wareham Market. Saw my Father there, his Wagon carried a Load of Wheat thither from Woodstreet for me. I Ballanced Accounts and so did my Father with Mr Cockram. Fine pleasant Wether. I just saw Father White there. I got home about 7. Talked with Amy Hunt who they say is with Child but she pretend 'tis some other disorder which she've had ever since the birth of her last Child. God knows.

28 May [Sunday]. I rode to Moreton Church in Forenoon. I saw Mr Lockyer and Was in and Eat and drank with him. I Met 15 or 16 of Wool Singers and went to Church with them. Text Matthew 11.29 first part and We Sung Psalms 149, 96^a, the communion Responsals, Psalms 21 and 48^a. Mr L[ockyer] Asked me to Stay and dine with him but I came home and dined with Father and J[oseph] K[ingston] and Joseph Davison and his Wife and they departed between 6 and 7. Father, Nanny, J[oseph] K[ingston], S[tephen] W[hite] and Joseph Davison went to Church. Text Philippians 2.12 last part and they sung Psalm 147. A fine Shower about 6, Blessed be God.

29 May. 5 Horses fetched 2 Load 1600 Brix and 3 Horses Rolled in Hoglease and Mountfield. The Masons arched over and finished the Oven. I took some of Dr Axford's Pills last night and this Morning which gave me 2 Stools. A fine pleasant day. Sowed Malt dust in the Wheat in the 6 Acres. Dairyman came and killed our fat Pig for us.

30 May. It rained, God be praised, Most of the Forenoon, a fine Moderate rain. 4 Horses about Middle of the Day fetched in fagots out of the Meads. About 40 Soliders Marched by here between 8 and 9 from Dorchester to Beer. About 4 Mr Willshear of Brockhole and George Woodrow's Wife came to mak my Wife a visit. About 5 My Landlord Mr Frampton came to See how the Building went on and Wished it had been built After my Plan. He goes to Bovington tomorrow to Sign Timber. A fine afternoon, very pleasent.

31 May. 5 Horses ploughed and broke up Robert Philps's little plot and 3 made an End Rolling for this year in Mountfield. William Bridle, a Stone Cutter of Blandford Who has Worked here some Time Rode my Mare to Blandford last night and this Morning brought a Letter For Mr Frampton which I carried to Bovington But He was gone home from Signing Timber and Father sent John Snelling To Moreton with the Letter. Nanny Rode with me to Bovington. There we Dined with my Father and Mother and rode from thence to Heffelton and there staid till about Sunset and Brought home my 3rd Son James Lyne with us. A little rain in the Morning but fine and dry Afterwards. 12 or more met at Church and sung and intend to Meet again next Friday night.

1 June. All the 8 horses with both carts and Farmer Talbott's [cart] which is yet left here behind carried dung from Dairy Barton into Mixon in long craft. About Noon Father White and Farmer William Burden called here as they came from Blandford with William Hodges of Worgret whom they took up and got a Licence to Marry him to his Sister-in-law *[space]* Allen who is with child by him. Just as they came Brother R[oger] and Mother White came from Heffelton and they all staid and dined here on Eels I sent some to Bovington this morning. When they departed Nanny rode with Mother and Brother R[oger] to Beer and is not yet returned. A dry day, Wind North East. Heads of the Kitchin Windows put on, Mr Fish came to see the Work.

2 June. All the 8 Horses with 3 Carts carried dung from Fords again and I rode over to Bovington and from thence with Father to Bindon and Woodstreet and so back through Wooll and Staid and drank at Mr Hayte's. I thought to dine with my Father and Mother but at Woolbridge met Daniel [Cobb] who told me that they had a Misfortune with a Mare, the Posstick slipped out and run into the Mare Vilots Kine just beside and was brok through into the inside so I got Dr Barrat to come with me and dress her and my Father came just as we had done her to see how 'twas with her. Father did not stay long. William White of Snelling [Farm] and *[space]* Hartnell of Winfrith came to see the Church Windows and Leads which are very bad. About 11 or 12 P[eople] met at Church to Sing. Very Cold in the Morning and most of the day.

3 June. 7 Horses with 3 carts Dung carted again and most cleared out the dairy barton. About Noon Mr Frampton came and staid a little While. I Went into Water and Bathed myself which I haven't done this 3 or 4 year before. About 3 Cousin Mary Porter of Ryehill and Cousin Hannah Fudge of Beer came to see us. I rode to Whitchurch and had a Young Puppy at John Galton's which they call Gamster. I hope my Mare which was hurt yesterday will do very Well again. A Hot day.

4 June [Sunday]. I and Nanny rode to Wool Church in Forenoon. Text Matthew 5.44 and we Sung Psalms 84, 48ᵃ, 26ᵃ and the comunion Responsals. My Father and Mother was at Church who with Many More I staid and received the B[lessed] Sacrement. We rode to Bovington and dined there with my Father and Mother and son J[oseph] K[ingston]. We came away about 2 and got home about 3 just soon enough to go to Church. Text Philippians 2.12 last part and they sung Psalm 43. After Prayer had a vestry about Amy Hunt who 'tis reported is with Child. She came hither in the Morning to convince my Wife [that she is not pregnant] But Nanny and all her neighbours think she is with child. Inclined to rain most of the Afternoon.

5 June. Small rain fell most of the Night and forenoon Mild foggy Air all day. Blessed be God this rain has been a fine refreshment to the Ground. John Whindle, Collarmaker, mended the Harness. Horses 5 with 2 Carts carried flints Out of Cockrode and Conigar into the Lane by Rawls's to make a footpath. Fetched 9 Bushels Grains from Beer at 3d. Inspected the Old Church Book. Mr Fish the Carpenter and John House his Mate came to Work so that I hope now the Building will go on apace.

54

6 June. A fine dewy Foggy Morning. Horses carried Heath and Gravel and made a good Way for the Cows to go out to Alderham down into Frogham and carried two Loads on the Flints we carried yesterday. After Dinner We Washed Sheep and I rode to Bovington and saw my Father [and] Mother and there was James Keat of Corf and his Son James they have been at Bindon cutting out Post Stones for Cow pens. Father was at Wool a Wheelbinding the BroadWheels and did not get home till about 9. I got home about 10. My Man R[obin] L[ocass] and I had some Words. He went to Coomb last Sunday and the Sunday before without my leave, which He said he would not ask on a Sunday to go home for his shirt. Began Weeding.

7 June. 5 Horses fetched 1600 Brix from Oakers Wood and 3 carried dung out of Cicilas Barton. I Was up with my Weeders in Morning. Mr Frampton came to see the Workmen and rode with me into Mill Pond consulted about Turning the Water there and [we] are to Meet about it again Next Friday. I rode Over to Bovington. Father was Sheep shearing with 6 men. I staid till about 3 Then came Home and with George Pain went to Beer with Amy Hunt supposed to be with child. There was a Vestry and the Officers gave us a Note to own she belonged to them and [they] would relieve her and her children, but no Mention is made of the Unborn child in case 'tis so. We did not get home till near 10.

8 June. 6 Horses fetched a Load of Lime from Middleburgh. Leonard Winsor and Richard Gallop of Beer came and shore my sheep which are 30 c[hilver] Hogs, 22 Ewes and 22 Lambs in all 74. I had a 2 yeared Heifer which could not Calve and I sent for the Dairyman and we had both hard Work to draw it, it was dead but I hope the Heifer will do well again. Fine Hot pleasent Weather. T. Hunt let the Water out of Gully and Turned up the Mud in inside of the Rails and with Charles Hodges and John and Simon Philps cut Hasssock Buts there. I think 'twill make a good Withy bed.

9 June. 6 Horses fetched another Load of Lime from Middleburgh and 2 fetched sand from Demerhill. Between 10 and 11 Father and Mr Frampton, William White of Snelling and John Lillington came and J. and T. Hunt with them to Mill Pond and Viewed it and consulted how to Turn it. Father came back and stayed till after dinner. Very Hot. The Great Wear at the Mill Pond is to be took up next Monday. We made an [end] Weeding of Salt Field where the Wheat is but Poorish. Between 6 and 7 Cousin William Fry of Beer with Farmer Stephen Ayles of High Town by Ringwood came to see us and stayed till about sunset. Mr Gold a shopkeeper of Beer came and Bought a Lean Pig of me and gave me 40s. for it but 'twas a good Pig.

10 June. I rode to Wareham Market. 5 Horses fetched the last Load of Frith from Bovington for this season and carried it into Gully, 3 Horses carried Gravel into Gully to a Sluice there and for the Wears at Lower end. I saw Father and Farmer White, Brother and Sister Florence at Wareham. Father and I came Home together so far as Gallows Hill. He told me that 2 of Mrs Sprat's Daughters of Woodsford Castle was at Heffelton that they was at Bovington and drank Tea last night and do intend to come and see us before they return. My Father told me somewhat of his Mind in regard to his Will and who He intended to Leave Trustees, JB. JG.* *[inserted in red ink]* James Lyne Died at Bovington 10 [June], 1757 aged 90.

11 June [Sunday]. I, George S[innick] and R[obin] L[ocass] went to Beer Church in Forenoon. Text James 1.11 and they sung Psalms 119, 28. I just called in at Cousin Fudge's and went to Church with him came home to dinner. I, Nanny, S[tephen] W[hite] and J[ames] L[yne] went to Church. Text the same as at Beer they sung Psalm 84. Cousin William Fry and Robert Porter were both here at Church and staid till near Sunset. Warm, Hot and a brisk Air.

12 June. A blessed rain from about 4 till 10 or After then 8 Horses carried dung From Dairy House to Long Craft with 3 Carts. 5 Men began digging out the Wear at Mill Pond. Mr Frampton rode round here in the Evening. Inspected the Poor Book. Sarah Brown of Wool came in the Evening to Work. Robert

* In fact, Joseph Warne appointed no trustees for his will; his son, James, was made sole executor. Thomas Bartlett (father) and Thomas Bartlett (son) were the witnesses. JB probably John Burt of Woodstreet, JG not known.

Duel Warne born June 12th 1757. He is blessed be God a fine Lusty Child but can't yet go. He being much Afflicted. The Other 3 all went at his Age.

13 June. 5 Horses fetched 2 Load 1600 Brix from Oakers Wood and 3 carried all the dung from Fords Orchard to Longcraft Mixon and the Men Dug out Mill Pond Wear and some of the New Main in Fursgoulds. Mr Frampton came just as they gave off and thought they went home to soon. The Esquire, his Lady and an Old Gentleman Rode by homeward about 8. A Fine pleasent day. Farmer Mate['s] Son of Beer came to see my Wooll. But I did not sell it.

14 June. Some rain again in the Morning. 4 Horses with the Wagon fetched Wood from Stockley for C. Hodges and 4 began carrying Dung from Home into Mixon in Cock rode. I and the Weeders in the Afternoon Made an [end] Weeding the Wheat, Cockled the Vetches and began in the Barly at Spear's. Mr Frampton Sent me a Letter by John Reason and Mr George, Servant to the Reverend Mr Portney, about the Land Tax Rate. Fine Evening.

15 June. All the 8 Horses with 3 carts carried dung from Home into Cockrode. I set out in the Morning to Bovington and saw My Mother and carried her a little Trout I caught last Night. Father was down at Bindon and I rode thither. He was on the return before I overtook Him. After I came Home went with the Weeders and done Long Craft and Bucks Hill and 9 or 10 Acres by C. Hodges's. Mr Frampton came and found fault with the People that they did not Work enough. Mary Woodrow and her Daughter Sarah, Farmer Bess's wife of Askerswell, came to see my Wife about 7 and departed at 9. Simon Philps and Joseph Boit began Mowing Ray Grass in Higher Hoglease, it is very thin.

16 June. Horses all at Dung cart as Yesterday I Went with the Weeders most of the Day and done Abraham's Plot Finished the Turnip Ground and began in Pit Close which is Very Thick of thistles. Mrs Ansty of Dewlish called here and Bought 4 Pigs about *[space]* Weeks Old at 6s Each. Coldish Brisk Air Morning and Night, Warm at Noon. Sent R[obin] L[ocass] to Bovington To desire Father to put the Timber Carriage in Order for us and he sent word they would send over the Timber.

17 June. 5 Horses fetched 2 Load of Brix and 3 with 2 Carts carried dung here at Home to Cockrode. R. Willshear, Farmer Talbut's man of Grange came between 3 and 4 and had away his Dung Cart. Between 6 and 7 My Father with his Timber carriage brought a Load of Timber 3 [?pair] Couples ready sawed out. Sent the Dairyman to Dewlish with the Pigs. Cut Weeds in Home plot, Conigar and Pit Close and some of us Haymaked in Hogs Lease. Mr Frampton's 7 Labourers began the New River course at Brandy Bridge, Opened it quite to the Mill Pond.

18 June [Sunday]. I and Nanny went to Beer Meeting. Text Ecclesiasticus 7.14 and they Sung Psalm 116, first Part. We was in at H. Bartlet's before we went in and sat in his Mother's Seat. We came Home to Dinner. William Locass came to see us before We went to Beer and just after We had Dined Brother Stephen White Whith *[sic]* Cousin William Fry. Brother S[tephen] had been at Beer Church and Dined with Cousin F[ry]. Cousin F[ry] Staid but a little and Went to Affpuddle and I, Brother S[tephen] W[hite] went to our Church [Turnerspuddle]. Text Philippians 2.12 last part and they sung Psalm 137. Brother S[tephen] W[hite] departed between 6 and 7. Close cloudy Wether for the most part.

19 June. 6 Horses fetched a Load of Lime from Midleburgh and the other 2 with 2 Carts fetched Sand. Some folk Made an end Weding below and done some in Mountfield. Some Hay Maked. Close Morning but very fine Afterward. The Waterman let the Water down through the New Channel but I think it must be Sunk much deeper. I rode to Bovington in the Evening and saw my Father and Mother and staid there till 8.

20 June. 4 Horses with 2 carts carried Gravel out of Fursgoulds Main which is Widened and deepened and began Making the New Dam or Bay therewith. The Other 4 Horses Dragged the little

Plot at R. Philips's. Took up Hay in Hoglease. Mr F[rampton] came to see how the waterman go on. He dined with Mr Plydell at Shitterton and called here as he went back about 8. He curses the Workman for a Parcel of Lazy Rogues. Very Fine Wether.

21 June.   Father called here between 4 and 5 as He was going to Ringwood. 4 Horses with 2 Carts carried gravel from the Main to the Dam and 4 Fetched a Load of Timber, 9 or 10 pieces, from Bovington, and After carried 2 Load of Turves to Beer. Dairyman and C. H[odges] Fired and cut Hassocks in Gully. Mr F[rampton] was here about Noon. William Barnes called here just before as He was going to Tolpuddle. 2 Tilers came to Work here from Dorchester. Fine Weather for the Hay. Masons Turned the Arch over the East Chamber Window.

22 June.   4 Horses carried Gravel again at the River and 4 [horses] and the Dairyman's Mare fetched 2 Load of Brix and Afterwards I got my Men and Weeders together and took up the rest of the Hay in Higher hoglease and carried it all together 12 little load of it, before Sunset. Father and Uncle Robert Duel of Totton rode by here just before I got home. They saw George and so did not call. A fine Afternoon.

23 June.   5 Horses carried Gravel at the River and the rest rested. We Turned the Hay in Horse Close. Made an [end] Weeding of the Corn and cut Thistles in Cockrode. Cut and Raked up Rushes in Gully. Nanny rode Over to Heffelton But Father and Mother White rode to Martin yesterday to see Brother and Sister Brown. In the Evening I rode Over to Bovington and Saw my Father, Mother and Uncle Robert Duell, got home between 9 and 10. The Masons Took down the South end Wall of the House down to the Eves.

24 June.   4 Horses carried 4 Load of Turves to Beer and 4 carried Gravel in the River. It rained a little so that We did nothing to the Hay. So I rode to Wareham Market at the River I met Mr F[rampton] and the Reverend Mr Milborn of Blandford with him, they came on foot and from the River up to the Dairyhouse and hither and staid here some time. Rose up the Couples and Pans and Fixed them. Wheat at Wareham Market from 5s. to 5s 6d. [a bushel]. I got home between 8 and 9. William White and George Paine had Amy Hunt to the Justices to Examine and remove her but coming home she was took in Travel and had a son born soon after she got home.

25 June [Sunday]. Farmer James Nipprod breakfasted with Nanny and Me, soon after my Father, Mother, Son J[oseph] K[ingston] and Uncle Robert Duel came. We went down round and saw the River, Cousin William Florence and another Lad with him and Uncle John White of Heffelton came to see us. Uncle R[obert] D[uel], Uncle J[ohn] W[hite], Cousin W[illiam] F[lorence] and his friend all departed soon after dinner and Father, Mother, I, Nanny, our 2 sons, J[oseph] K[ingston] and S[tephen] W[hite], all went to Church together. Text Philippians 2.12, last part. About 6 Father and Son J[oseph] K[ingston] departed and my Mother staid here. My Servant R[obin] L[ocass] went home again this Morning without my leave and I Whipped him when He came home. He is very Impudent and Saucy.

26 June. I Took a Great Cold lately and had a Soar Throt which I found something of last Night. 4 Horses carried more Gravel at the River and 4 carried 3 Load of Turves to Beer and Afterward 5 Load of Hay out of Horse Close. My Man or Boy R[obin] L[ocass] was so affronted last Night that he asked me if I would pay him and he would be gone, for he would not bide to be Beat (see the 6th). But on my reprimanding him He took it into Consideration and went to Cart as usual. Mr Frampton was here about Noon. I carried home my Mother in the Evening as I returned I Met Dr James Going to Bovington.

27 June. 5 Horses fetched 2 Load of Brix and 1 Cart fetched Laths from Tolpuddle. Afterward fetched in the rest (2 Load) of Hay out of Horse Close and put the Cows in there at Night. Cut and raked rushes in Gully. Mr F[rampton] was round here to Day but I did not Meet with him to speak with him. The Tiler Thomas *[space]* began covering in the house at the West-South-West Corner. Had a swarm [of bees] The 1st this year. Dry Wether Very hot.

28 June. I and George [Sinnick] with 4 Horses went and carried a Load of Turves to Bovington and brought back a Load of Timber for the Lower Great Wear to be set up in the new Bay. 4 Horses with 2 Carts Carried more Gravel to the New Bay. Mowers Made an End cutting Brick Close and began in South 7 Acres. After the wagon came from Bovington it carried 2 Load of Turves to Beer. Father and Mother White came about 4 and departed about Sunset. Very Warm.

29 June. 4 Horses and 2 Carts carried Clots and Gravel to the Dam and a Bay to Set the Wear. Simon Hallet, Robert Crocker and another man are fitting it up to set it tomorrow. John Allen is very Bad, He've just been here to look on them. I and George [Sinnick] with 4 Horses fetched a Load of Brix from Oakers Wood Kiln and since I've sent Daniel [Cobb] with him to Beer with 2 Load of Turves. Mr Lockyer, Bailif to Mr Frampton at Moreton, Died Suddenly this Morning. We Pooked up Brick close and fetched in 5 Load of it. Mowers Finished South 7 Acres and began in Lillington's Mead. Good Weather. Masons Finished the New Chimny and the Tilers began covering the South side of the House. Mr Read another Tiler came and his Cousin Read which last, because there was no Beer Allowed, went home again.

30 June. 4 Horses Fetched in 3 Load, the rest of the Hay in Brickclose and afterwards with 6[horses] carried Mud out of Street up into Cockroad and 2 Horses rested. I rode over to Bovington and to Wool and Bindon where my Father was Haymaking and I brought home a Scythe that was at Woodstreet, Brother R[oger] White was here about 11 after Barm. The Great Wear at Mill pond Fell down on R. Crocker, Charles Hodges and J. Lillington and very much hurt R[obert] C[rocker] that He rode Home. A little rain in the Morning but fair Afterward. Mowers Finished Lillington's Mead.

1 July. 8 Horses with both carts carried Mud out of Street and Gully to the Mixon in Cockroad. Mr F[rampton] came about Noon and saw the People. Farmer Midway came soon after He

[Mr Frampton] was gone and Wanted to speak with Mr F[rampton]. 'Twas (as he told me himself) to ask Mr F[rampton] for his Bailiship for his Son John but I don't think that will do. It rained a Shower or 2 in the Morning and Afterward No Haymaking. The Folk began setting in the 4 Hatched Wear at the River. I rode about Heath in the Evening to see for 6 Heifers which are stragled away, but I could not find 'em. Mr F[rampton]'s Coachman came in the Evening to envite me to Mr Lockyer's Funeral tomorrow Morning, 8 o'Clock.

2 July [Sunday]. Between 7 and 8 I rode to Moreton and there met my Father and Several others and from Moreton Attended Mr Lockyer's Corps to Hampreston. There was 25 Horses and 25 Men Set out together with him from Moreton and not one Woman. We had more company Joyned us at the World's end and Wimborn. We got to Hampreston about 3 and He was buried before 4. My Father, I and Edward Hooper of Moreton was in at Farmer Hayward's and at Wimborn We staid and fed our Horses and We only came together to my Gate. 'Twas between 9 and 10 when I got home. Nanny was at Church. Text Proverbs 20.19 and most People said 'twas a Very Odd Sermon. Nanny did not like it. John Johnson, I and Farmer Threisher on the Right, Farmer Antrem, John Saunders and Farmer Shutler on the left supported the Pall. A fine Rain Blessed be God.

3 July. 6 Horses fetched a Load of Lime from Morden Kiln And Father's Wagon fetched another Load for Us, and 2 Horses fetched sand. Fine Wether for the Hay. Father's and My Carters was all to much concerned in Drink. Masons Finished the Brewhouse North Wall and began pulling down the Old porch or Stair Case.

4 July. I and George [Sinnick] Fetched a Load of Stone with 5 Horses from Wareham for Tiling. 3 [horses] with 2 Carts carried out Mud in Mill pond and Some Dung. Afterward they carried in the Clover hay in 7 Acres. Cousin Joseph Warne and Farmer Stephen Ayles called here as they was going to Sheep shear and I rode with them to Bovington and Father being Haymaking at Bindon, We rode thither to him and up to Woodstreet and back to Bovington to supper. And S. Young continuing Ill and The

Dairyman at Bovington not agreeing with his Daughter-in-law, Betty [Oliver], she came hither with me till S. Y[oung] is better.

5 July. 4 Horses Fetched a Load of Brix and 4 with 2 Carts made a way into Lillington's Mead, Afterward carried hay from thence *[space]* Loads. I Rode to Dorchester fair and drove to Jumping Heifers and sold 'em for £6 and bought a black Mare Colt 2 years old for 7 Guineas. Sheep and Lambs sold Dear, Cow Cattle not so dear as it have [been]. I saw Cousin Joseph Warne and Farmer S[tephen] Ayles there the Hampshire men sold their Wool at Winfrit for 22s 6d. a weight. A Fine day got home about sunset.

6 July. I and George [Sinnick] with 4 Horses and the Colt fetched another Load of Tiling Stone from Wareham and 4 [horses] carried Gravel to the Wear and Afterward carried 4 Load of Hay, Viz. the rest of Lillington's Mead 2 Load and South Mill plot 2 Load. Mr F[rampton] was here Tuesday and today got up the Roof of the Brewhouse. Tilers done the East end of the New House. Fine Weather. Carpenters made a Step Ladder to go to Bed 2 nights before We Went out round and up by a rong Ladder. Nanny did not like it.

7 July. 7 Horses carried Hay all day made a little Reek in Ford's Cowlease for sheep Cleared the North Mill plott, Cicilas Mead and some of the Moor. About 4 in the Afternoon Sister [Elizabeth] Loxly came, she came from Martin yesterday to Whitechurch in a Cart and there laid at Galton's and He brought her hither, a blessed day. Finished a Reek at Fords and began another.

8 July. 5 Horses fetched 2 Load of Brix, a Load of Fus fagots for Bedding and a Load of Hay and 3 Horses Fetched in 3 Load Hay, Viz: 2 out of Moor and 2 from Robert Philps's Farm for which I am to give 35s. a year. The Dry Hay all in. Wether seem Inclined for rain. Sister Loxly departed about Noon for Bovington.

9 July [Sunday]. I and Nanny went to Beer Meeting, Text Romans 5.2, last part. They sang Hymn *[space]*. Cousin William Fry and my Father *[illegible]* went to Church here, Text Luke 11.13 and they Sung Psalm 66th. Dairyman's Brother at

Bovington came hither about 5. They Want to get back Betty again, but she don't seem Willing to go. Father and He departed between 6 and 7. My man George [Sinnick] out last Sunday Night and this Night and did not come in to Supper. Some small rain about 10, dry afterward.

10 July. 6 Horses Fetched a Load of Stone from Wareham and 3 Plowed R. Philps's Plott again for Turnips. I rode to Blandford Fair. I had a thought of Buying some Lambs but did not. They sold very dear and Went Off very soon. I saw Brother Brown there and came home with Father White. Rain about Middle of the Day and came on Brisk in the Evening. William Hodges came in the Morning for the Dairy Maid, Betty Oliver, her Father-in-law wants to get her back home again.

11 July. 6 Horses drew the 2 Wagons to Wool where We Staid and had 1 of them Mended and left the Other there so with the 6 Horses We brought home some Spar Gads, 4 sacks Wheat, all that I had there, a Grindingstone, a large Tub and the Fan Tackle, a Table, the Furnace and Other Lumber Things. I saw my Father at Wooll. Dined at Woodstreet with T[homas] Barratt and his Wife. I called at Bovington as I came home and staid there till 6. We found our Young Heifers as We Went. A Smart Shower of Rain from 3 till 5 in the Afternoon.

12 July. 6 Horses fetched another Load and half [of] Stone from Wareham, this Makes 6 Load which We've fetched at 4 Times. 3 Horses carried out Mull from the Reek yard into Street and some dung up to the Mixon. It rained very fast about Midnight, the Hay was very Wet. We turned some of it the Afternoon. Mr F[rampton] was here today. He thinks the building goes on Very Slow. A dry day but the Wind brisk.

13 July. 6 Horses carried dung [in] the Forenoon to the Mixon. I and Daniel [Cobb] got in the 6 Heifers out of Heath which have been missing Some Time and 4 yearlings and 3 I could not find. Much rain in the Night and Morning, Dry Afterward. Haymaked in Afternoon. I Ordered my Man Daniel [Cobb] to help Milk the Cows and He said he would not and went out of the Way and did not but, being come [back] before they was done, I bid him again and on his refusing it laid him on with a little hazzle stick and he got another but did not Strike me. Mr Fish and Mr Garrard parted us. Took down the North Gable End.

14 July. 4 Horses and the Colt, George and Simon Philps Went to Wool and fetched home the Wagon left there last Tuesday and brought home Some Clift wood from Woodstreet and the rest of the Fan Tackle. 4 Horses with both Carts carried dung up into Cockroad. Afterward carried 7 Load of Hay out of Rack Mead. Drying day. Gave the Haymakers some Ale. We Baked in the New Oven the first Time and it baked very Well. Nanny gave all the Work folk about the building some Cake and Ale. Daniel Told me He would not Milky again. I told him if He did not He should not Eat, nor would not let him but Thrust him out of door and footed his Arse. He seemed displeased at that and would go away if I would pay him but, on my Threatning to have him before a Justice, He Went and Milked and Made no more ado about it. Took out the North Chamber Windo.

15 July. Horses (6) carried dung out of the Reek yard into Mixon in the Morning, but the Rain beat 'em off, it rained most all last Night and the Wind High. Mr F[rampton] Was here Yesterday and Ordered that Mr Jordan of Blandford should make the Casements for the Windos here and to get Stone for the Kitchin Floor. So I rode to Rye hill had my Horse shod, went to Beer and paid my Collar maker and rode to Wareham and so to Corfe and there bespoke Pave Stone of James Keat. I staid a while at Mr Cockram's with Mrs Jeame from Wareham with Father White.

16 July [Sunday]. Done nothing to the Hay yesterday. A Shower or 2 in the Morning, Fine Afterward. After dinner Cousin Jonathan Fudge of Beer and his Wife came and Went to Church with us. Text Luke 11.13 and Beer folk Sung Psalm 28. Mr Fisher's Maid Charity with Farmer Spear's Wife and Children and William Rawls's Wife and son, Pat, Trim and 2 of Mr Moors's Children and Farmer Spear was all in here before they went to Church. Cousin Fudge Staid afterward till 8 Clock. Amy Hunt's Basterd baptized John.

17 July.   6 Horses carried dung in Morning from the Reek yard into Cockroad Mixon, Afterward the rest of the Hay out of Rack Mead *[space]* Load there was in all *[space]* Loads and they carried 1 Load out of Horner's Mead. I rode to Bovington in the Morning did not see Father for He was gone to Bindon a Haymaking. 'Twas a blessed day. Brother Loxly came to Bovington yesterday and He and Sister came from Bovington a little after me and departed hence about 3. Afterward Nanny rode to Heffelton, got home between 9 and 10. Mowers went on again having stopped Some Time.

18 July.   6 Horses fetched in the rest of Horner's Mead and what was cut before this Week in Farm Mead was standed Twice or More. The Way out is very bad. Began a Reek here at home behind the Stable. Yesterday the Masons finished the North Windo. Today pegged out the Brew house After 'twas Levelled and got stone in order to lay the Floor. The Carpenters Yesterday got of[f] part of the Old Roof at the North End and got up the new Hips and began the Stairs today. Mr Frampton was here yesterday and today I told him Cousin William Fry had a mind to buy William Alner's Living in case he would let him buy further but that Mr F[rampton] don't choose to do.

19 July.   6 Horses fetched a Load 900 of Brix and then left 1 Horse and took the Colt and fetched a Load of Pave Stone from Wareham 140 ft. and 3 Horses fetched Sand and dug out a way in Mead. A fine Morning but about 10 or 11 Some rain, several Showers Afterward and some Thunder at a distance and much rain in some places but we had very little here. The Carpenters laid Steps of the New Stairs and Masons 4 courses of Brix and above 2 of Stone in the Brewhouse. Old Farmer Alner came and saw us. Since 5 My Wife went and saw Mrs Willshear and is just now returned. Cousin William Fry was here this Evening.

20 July.   With 4 Horses fetched a Load of Heth into Mead and Afterward with 4 more carried dung and Rubble up into Cockroad and some Brick bats to make a good way into Mead. Mr F[rampton] was here and liked our Way of doing it very Well. I and T. Talbot made a Screen in the Morning to Screen the

66

Rubble and it Answers very Well. A blessed day for the Hay. We had a Misfortune and hurt the Gray Mare Vilate again. She was next to the forehorse which Turned about because of the Flies and she stepped over his Traices and the Spreader ran up into her Groin 3 or 4 Inches. I sent to Wool for Dr Barratt who came and dressed it. John Elliott also called here. All the Steps of the Stairs which go into the new parts laid.

21 July. 5 Horses with 2 carts carried gravel to make Fords in the Mead till about Noon. Afterward was going to carry Hay but it rained apace for some Time and We got in but part of a Load. Between 3 and 4 I and Nanny rode to Bovington and Brother S[tephen] White came the While to See us and rode from hence to Bovington there We staid together till after 8 and got home just before 9. Father killed a great Bull this Morning near 30 score and cut it up while We was there and gave us some of it.

22 July. I and George [Sinnick] with 6 Horses fetched a Load of Stones and Father's Wagon was There and brought another Load for us. I Staid a While behind the Wagons but got home as Soon as they. Fine Wether for the Hay. The Other 2 Horses carried away Rubble from before the House into Conigar and before We came home the Dairyman got his Own Mare and borrowed a Wagon and Horses of Farmer Alner and got in 2 Load of Hay and We carried afterwards till Sunset 8 Load in all. [?Put] 4 in a little Reek for the Sheep in Brick close. I and Nanny had Words about a Ring of Keys which she let the Children loose 'Twas before the Workfolk whom I wanted to pay. I Was sorry for it Afterward. Wheat at Wareham Market from 5s. 2d. to 5s. 9d. [a bushel].

23 July [Sunday]. Coldish Wind and Cloudy Wether. I went to Beer Church in Forenoon. Text Proverbs 16. 29 and they Sung Psalm 119 and Revelations 19. I and Nanny Went to Church in Afternoon but She did not like the Sermon, she Talked of going to Beer Meeting had I been in the Way. Text the same as at Beer. They Sung Psalm 95.

24 July. 6 Horses fetched the last Load of Floor Stone from Wareham and coming up Beer Heath broke the Fore Axle so unladed into another Wagon and brought all home. Mr Keat and his Son James, the Stone cutters, came. Several Showers in which abundance of Rain fell some Thunder to the North. No Haymaking. The Stone Cutters who would feign have Lodged and dieted with us do Quarter at the Dairyhouse. I rode to Beer in the Evening. Mr F[rampton] was here was not pleased because the Tilers was not come. Masons began setting the Furnace Saturday and finished it this day. Mr Fish put Studs to part the Chambers and the Brewhouse inside door by the Milk house.

25 July. I rode to Wool in the Morning for J[ohn] Allen to come and Ex[amine] my Wagon again and rode down to Bindon where my Father was with his Wagon and Men Mending the Ways and I helped 'em in with a Trunk as one go up from the House to the South Barn. I called at Heffelton as I Went along and Staid about ½ an hour. 5 Horses with 2 Carts carried away more Rubble and dung. A fine Morning but since I came home another Shower so no dry hay, yet a very Fine Evening Afterward. The Garret Stairs begun and the Kitching Floor began laying 4 Stones Masons Stopped the Scaffold holes and trimmed the Windos.

26 July. Carried 4 Load of Gravel with both Carts to make Ways in Mead. Afterward it being dry though Cloudy We fetched in 6 Load of Hay but the rain came on about 2 and Beat us Off. Between 5 and 6 my Father came to see us but did not Stay long. Masons put in the Windo Bars the Stair and Milk house Windo lights. Put up the Stone Cutters diet here for the dairy Woman did not care to Take the Trouble of it but they Lodge there. The Garret Stairs up to the last Step. Mower[s] finished cutting the Grass yesterday.

27 July. A fine drying day. 5 Horses carried away Rubble into Cockrode. Afterward We borrowed Farmer Alner's Wagon and with my Own 2 We fetched in 10 Load more of Hay. Stairs finished and Some of the Garrat Floor laid. Masons began Lathing between the Chambers to part[ition] it. The Tilers finished covering the House but has some Old to point. Nanny Rode to Beer and had a Tooth Out. Sister Patty came soon after and rode to Beer After her and from Beer home.

28 July.   Horses carried away some Rubble from before the Door. It rained Early in the Morning and Much fell in the Forenoon a fine Afternoon. We got out and turned over a little Hay in the Evening. Old Mr Rogers the Tiler and Robert Philps fell down both in the Ladders from the Gutter in the North East part of the House. Lathed the partition between the Chambers.

29 July.   A dry Night and blessed be God a Fine day so We finished Haymaking and should have fetched it all in but We fetched a Load out of Brince Puddle for Farmer Alner and with the last 2 Load We got out this Night We Was very Much Standed so that We did not sup till 10 or After and We left out 2 Little Loads More, which Else we could have got in Very Well. Masons Plaistered between the Chambers, the Floor of the Garret pretty Well Laid. Mr F[rampton] came but I was in Mead and did not see him.

30 July [Sunday].   I, Mr Keat and his Son James Went all together to Beer Church in Forenoon. Text John 9.4 last part and they Sung Psalms 81, 5. We Went to our Own Church in Afternoon. Text the Same as at Beer and they Sung Psalm 25. A List of the Militia was put up to the Church Door. There are 16 in our list. A Fine day. The Reverend Mr F[isher] preached an excellent sermon this day.

31 July.   We fetched in the rest of the Hay, 2 Load, and Afterward carried some Spear and dung and Ashes from dairy-house. I and Nanny thought to have went to Heffelton but Lucy Willshear came in so Went not but I rode to Bovington. Father is very busie about his Hay. I Offered to send a wagon and help but he did not choose it. I got home between 8 and 9. We trimed and Tiped all our Hayreeks. A blessed fine Afternoon.

1 August.   3 Horses Ploughed and 4 with 2 Carts carried Clots out and red Earth into R. Philps's Plott which I think to Sow to Turnips. Mr Metyard came about 1 and Measured the Tilers' Work. He dined with us and soon set out for Wareham and I and Nanny rode to Heffelton and staid there till 9. Mr F[rampton] came after we Was gone and Was very angry with the Workman because They had so Floored the Garret that the Chambers under must be ceiled, which He said a Saturday should not be done but now consents to it.

2 August. 3 Horses Made End plowing and Harrowed R. P[hilps]'s Plott and 4 carried Earth and Ashes enough to do it Over and so if God permit I'll Sow it to Turnips tomorrow. Nanny rode to Rye hill to See If Uncle John White was at the Duke's head for he have been Idleing about ever since last Sunday. While she was agone Brother Michael Saunders and his Wife came about 6 o'Clock, they Came from Martin in the Morning and rested a while at Galton's at Whitchurch. Fine Weather.

3 August. 4 Horses I sent with the Wagon to Heffelton to Hay Cart and the Other fetched 1 Thousand Laths from Tolpuddle and got up Spear. Sowed and Harrowed in Turnips at R. P[hilps]'s and carried out the Clotts so that it looks quite fine. Mr F[rampton] came about 11 and was here an hour or More. I Asked him to build a Necessary House, which He consents to as also to repairing the Skilling at dairyhouse. Brother Michael S[aunders], his Wife and Nanny with them Set out about 1 for Bovington. I rode to Bovington to come home with Nanny but met her at Conigar Gate and so rode no further.

4 August. I Awoke soon after 2 and by 3 had a Violent pain in my loins at the left side. I could not lie a bed nor was not Easy up, it held me about of an Hour and then, blessed be God, I grew Easier but had Such another pain again about 2 or 3 this Afternoon but I thank God it did not hold me quite so long. I began Making a Will. 'Twas a blessed fine Forenoon but about 4 it rained. My Man, Daniel Cobb, Set out from me this Morning, but I paid him no Wages, so I Wot he'll fetch a Warrent for me. 'Twas because I found fault with him for penning the Sheep so Soon last Night for 'twas not 7 by my Watch when He penned them. Met about the Militia at Wareham. Took down the Old Chamber doors and laid the Stairs.

5 August. 6 Horses carried away Rubble from before the House into Cockroad. After Dinner I rode to Heffelton, Wool and

Bovington and Brother M[ichael] Saunders and his Wife came from Bovington hither before me. Fine Weather for the Hay. Wednesday *[2 August]* : Stone cutters finished Laying the Kitchin floor. Thursday: Laid that by the Stairs and Opened the door Way out of Kitchin into Parlor and laid it and yesterday Old James Keat went home and Y[oung] James began laying the Passage. Mr Fish Finished the Old Chamber Stairs but not put up the Doors. I Took Daffy's Elixir which worked pretty Well.

6 August [Sunday]. I took Physick again Which Worked 5 or 6 Times. Brother M[ichael] S[aunders] and his Wife departed about 9 for Martin my Father came over in the Afternoon with my son J[oseph] K[ingston] and Went to Church together. Text Psalm 58, last part. My Father departed about 6. At Church they Sung Psalm 147. We had a Consult about Swearing Amy Hunt to the Father of her Child which is to be done this Week. George Paine, Farmer Joseph Alner and Farmer Willshear was with her in the Evening about it.

7 August. 6 Horses fetched some Deal Boards and some Stone from Wareham and 3 carried away rubble into Conigar. I went with the Wagon and Mr Keat and he came again in the Evening. When I came home Nanny was very Sick, she was took soon after the Morning. Picked up Aples at Cicilas Orchard and began Cutting Peas in 5 Acres. Masons plaistered the Garret. Mr Fish most drunk.

8 August. 3 Horses fetched in 2 Load Fusfagots and 3 carried away Rubbish and Afterward fetched a Load 800 of Brix from Oakers wood. Mr Keat and his Son finished laying down the Passage Floor and Stones before the Doors, which in all is 484 Feet at 4½ d. comes to £9 1s. 6d. Finished cutting Pease and began Cleaning away the Rubbish in the Reekyard which is to be a Garden. Some rain in the Morning but pretty good Harvest Weather Afterward. I was at George Woodrow's and they made me very Welcome.

9 August. 5 Horses and 2 Carts carried away Rubbish Out of Reekyard up into Cockroad. We picked up Aples, Turned Pease and Rushes and Cleaned Brick. Joyners hung up the 2 Old Chamber Doors. Mason stopped holes in the Ceiling of the Old Chamber, cleaned down some Rubbish out of the New Garret. Mr Keat and his Son departed this Morning After I paid him his Bill which was £9 3s. 2d. A fine Forenoon, a Shower between 5 and 6.

10 August. 5 Horses and 2 Carts carried away more Rubbish out of Reekyard, some into Cockroad and some into Hilliar's Plott. Several Showers of Rain, but more I believe in some parts than here. Between 1 and 2 some Very Loud Thunder Eastward. The Parlor door put up and Quarters between the Parlor and Cellar Durns to the Milkhouse and 2 cellars. The Masons plaistered the Portal going into the Old Chambers and partition between the Parlor and Cellar. I and 5 Women cleaned some of the Old Brick.

11 August. About 9 George [Sinnick] and J[ack] Hodges set out with the Wagon to Blashknowle beyond Corfe and fetched a Load of Lime. The other 3 Horses fetched 5 Load of Earth and 2 of Chalk and the Masons began to Mud wall up the Gate way (leaving only a door Way) between the lower Barton and the Reekyard. Joyners took down the Old partition between the Milkhouse and the Old back passage which with part of the Milkhouse is to be made into a Cellar and they set it up again further in. I rode to Bovington and saw my Father and Mother and brought home a Chain, an Iron dog, some Barm and Brocoli plants. Father began Wheat Reap yesterday, had 26 [?bushels] this day. Fine day.

12 August. 6 Horses carried away Rubbish out of Reekyard into 4 Acres, carried Spear to Cicilas and got in all the Peas, 4 Load. I rode to Wareham Market about the Windo Rate. There was Amy Hunt and she swore Joseph Alner was the Father of her Child. She was very loath to Sware and begged of me and of Father White for Christ's Sake to be Excused a fortnight longer and Wept bitterly, but she could not prevail with us. Her fear was that the Child would be took from her and sent to the foundling Hospital. Masons Wetted up the Load of Lime and set in and began Walling up Durns between the South East corner of the House and the Cart house. Joyners did somewhat to the Partitions of the Cellar and put 2 doors in the Kitchin.

13 August [Sunday].   Smal drisling rain most of the Forenoon. I and Nanny Went to Beer Meeting in Forenoon. Text Acts 26 and they sung Psalm 119. After Dinner Betty Porter, who went home last night, came again and We Went to Church, Text Romans 12.10 and they Sung Psalm 4. Joseph Alner was at Church but We had no Vestry about his Bastard. Some say Amey [Hunt] is set out to Beg the Country with it rather than it should be sent to the foundling Hospital.

14 August.   6 Horses carried away Old Mudwall from Reekyard to Hilliar's Plot. I had 5 Men began wheat Reap in Saltfield. We Hay Maked in Pin Mead and in the Evening I got in 6 of the yearlings. [Masons] done up the Wall between the Great Cellar and Milkhouse and 1 pane by the Passage and beat up the Old Floor and began new laying it. Joyners hung the Milkhouse door, made Chamber doors and Cellar Windo frames. Mr Fish Went to Moreton. A blessed fine day for Harvest. I've let my Wheat to be cut by the Acre at 4s.

15 August.   Some rain in the Night and Much in the Morning. 6 Horses fetched a Load 800 of Brix and Afterward drugged up Elms which I had routed at Hilliar's last Michaelmas out of which I think to rise Timber for a Reek Stavel. The Masons finished the Great Cellar Floor, Walled up the Old back door and put in a Windo in its Stead. Joyners put up the Garret and both New Chambers Doors and done a little to the Clossett. We fetched in One Bed out of Barn and put up in the New Garret. C. Hodges set Brocoli plants. A dry Afternoon, reaped some.

16 August.   Much rain in the Night and more this Morning. We fetched in the Other bed out of Barn and set it up and put up the Deals over the Porch. Horses fetched Sand and carried Clots to make up Bays to turn the Water into Cicilas Mead. I, C[harles] H[odges] etc., set abroad Wheat Sheaves and turned Oats. Masons put a Windo into the little Cellar and plaistered partition and Walled it. Joyners Finished the Stairs, railing them (Mr Fish departed and left John House and John Bonnell to Finish which they intend to do a Saturday) and Boarded over the Portal into the Old Chamber. Fine Weather. Fetched in 2 Load of Wheat.

17 August. Horses carried away Mudwall out of Reek yard into Hilliar's and druged up the rest of the Elms out of Lane to Sawpit. Afterward We fetched in 5 Load of Wheat all that was cut in Saltfield and Got in all the Hay (2 Load) out of Pin Mead. I and C[harles] H[odges] turned the Water into Cicilas Mead, Cleaned the End of the Barn and Bedded it with Deal Shavens. Masons finished the Cellar Wall and beat down the Old Wall and Ceiled the Passage and plaistered the Cellar Wall and Walled up the Old North Chamber Windo Seat. Joyners Boarded it put a door to the little Cellar and between the Garrets and the 2 Chamber clossett. The Fleas are Busy about the Sheep 'tis difficult to keep 'em free of Magots.

18 August. Some Rain in the Morning but not much, but I did not let my Men Reap, but they dug out fords in Mead the While the Horses fetched out Spear. And Afterward some of them filled Rubble Brick in the Reekyard and we carried it pretty well all away down into the Mead and Afterward we Went and pooked some Oats in Long Craft. Masons Lathed the best Chamber Closset and Walled up and set plugs in the Windos and Ceiled the 2nd time the best New Chamber and the South Wall. Joyners put in Windo Seats in the Chambers and Kitchin, hung the Bottle house and great Cellar Doors.

19 August. Horses fetched up the last Load of Spear out of Mead and Afterward all the Oats in Long Craft, 5 Acres and Buckshill which was 10 Loads but 'twas late when We got in the last. I and 5 Women Pooked 'em. Masons Finished Plaistering the best Chamber except the Closet. Joyners cased up the Windo, nailed Strips over the Portal doors, put a Lock on the Great Cellar door and done Privy House door durns and about 3 departed and Carried off their Tools with them, Going next to Rushmore Lodge in Cranborn Chace. Blessed Fine Harvest Weather. J[ohn] Allen's Men, 3 of them, about a Reek Stavel for me.

20 August [Sunday]. I and Nanny Went to Church in Afternoon. Text Proverbs 25.12 and they Sung Psalm 103. Farmer Dolling of Southover by Tolpuddle was at Church and in here after. I and Nanny to go with him to George Woodrow's where his Wife was

74

and We was Going with him and was beyond Droop and George Called after me For Brother Roger and Sister Patty White was come they had been at Beer Meeting. Text there 1 John 5. 3. They Staid till Sun set. Father's Men, Sillaway and Hooper, was here at Church and told me that Mrs Robinson was at Bovington, that she came the Thursday.

21 August. 5 Horses with both Carts carried Rubble out of Reekyard into Hilliar's Plot and Conigar. There must be more carried away yet. Fine Weather. R. Philps and John Gould made an End Reap[ing] in the 6 Acres and Bound some of it. The Leasers in Salt Field Quite Rude. I rode to Ryehill and Spoke with W[illiam] Welch about binding Dung pot Wheels. Masons Ceiled the little Chamber and plaistered the East side and finished the Closet.

22 August. We fetched in the last of our Wheat, blessed be God, in Very Good Order. They finished cutting it but just in the Evening I made a little Reek of 6 Load. We had in all Saltfield but 15 little Load and but 3 little Load in the Ground at Fords next to William Alner's called 6 Acres. Masons finished the little Chamber. J[ohn] Allen's Men finished the Reek Stavel, Mended the Great Barns door and Durn and nailed up Rails between Moor and Buckshill. Fine Weather Continues. I was at Beer in the Evening to change some Gold for Silver and Mr Vander[plank] told me and Mr Bartlet was at Bovington Sunday Evening and came home Monday that Mrs Robinson was there then. Mr R[obinson] and she have had a fall out lately but I can't think how she come to take refuge at Bovington.

23 August. We ploughed up the Reekyard which is a bed of Gravel, especially the lower part so We are carrying it away and makes a Road therewith in the Mead. In the Evening I and Robert Phillips measured the Ground called 6 Acres which He, his Wife and John Gould reaped and it is but 4a. 3r. 28p. and R[obert] P[hillips] was very uneasy that I did not pay him for the full 6 Acres. Masons Almost finished the Stair Case. J[ohn] A[llen]'s Men Parted the Great Mead and hung the lower Garden door.

24 August. 5 Horses and 2 Carts, 1 Man to drive and 3 to fill, carried Gravel again out of Reekyard into Mead. I Pitched the Fold as I do most Mornings, looked to the Sheep and Moved Cattle, put the Milch Cows to Yea Grass in the South part of the Mead. S[arah] Brown from Wool came this Morning to work for Nanny. The Masons finished the Stair Case and began plaistering the Kitchin. I and Nanny rode to Heffelton about 3 and returned about 8. Father White and Sister Florence was gone to Handley to see Sister Brown who with her Children was carried a 2nd Time from Martin with an Order and their biggest Child Jacky 'twas Said was taken Ill of the small Pox. Mr White did Weep very much about it. Several showers of Rain in the Evening.

25 August. 5 Horses carried more Gravel and Stones out of Reekyard down Into Mead. We plowed the Garden again. I had Women to help pick out the Stones but the rain beat 'em off. After Dinner I rode over to Bovington. My Father was gone to Woodstreet and did not get back till about 7. I staid a little After he came. Mother Told Me Mrs Robinson went home Tuesday and Father carried her [i.e. mother] down to [the Robinson's] in the Evening, that Mr Robinson Beat Mrs in the Evening [and] that She [mother] stood up and thought to have parted 'em but they with the Table was down about house all together, that Mr R[obinson] was set out the Next Morning before they upped and that Mother thinks she shan't go thither again yet in an Hurry. My Mother charged me not to tell it again to any body.

26 August. With 4 Horses and 2 Carts We carried away [a] Load Stones into Mead and Afterward the Top Earth though pretty Gravelly into Gully. Between 1 and 2 Shut out the 4 Horses, Shut on 4 Fresh Horses and carried till 4 Clock then the Wether being Cloudy in the Morning but dry We Turned and got in all the Vetches and We Turned Barley in the Evening. Masons Plaistered yesterday and this day in the Kitchin and Finished it.

27 August [Sunday]. Nanny longed to hear how Sister Brown's Child was so I rode to Heffelton and 'tis not the Small Pox and the Child is pretty Well again. I did not stay but rode to Stoke Church, Text Romans 2.4 and they sung 100ᵃ, 92, 139 Psalms. I was just in at Mr Rob[inson's] After we came out of Church they would have had me staid but I would not. I Talked with Mr Robinson and his Wife and told 'em there [sic] way of Life was very Scandelous and Wicked, they exclaimed very much at Each Other and I found talking was like to do more harm than good so I rode with Mr R[obinson] to Wool Church. Text the same as at S[toke] and We sung Psalms 66, 42, 103. My Father and Mother was gone to Moreton so I did not see them. I came home just soon enough to go to Church, Text Job 8.13 last part and they sung Psalm [space]. My son J[oseph] K[ingston] with John Sillaway was here and Richard Duel and his Wife and they all departed soon after Prayer. Miss Ekins was in here after Prayer and staid till about Sunset. I and Nanny bore her Comp[any] to Chamberlain's. Some Showers.

28 August. 5 Horses fetched a Load of Lime from Middleburgh and 4 fetched Sand, Chalk and Earth for Mudwall. Much rain before and After Sunrise but Slacked about 10, then I and Nanny rode to Weymouth to see the Great Fleet which now lies there in the Road there are above 100 [ships]. They have made a decent [sic] twice this summer on France and brought of plunder with them which they sell there but the Wind was so high that 'twas very difficult to go out on bard them, yet some did venter but We did not. We went and saw Captain Weston's Wife and Sister and her Sister Miss Atty Jeanes. We dined there and they walked out round the Town with us and Afterward drank Tea with 'em. They was Exceeding kind and Civil to us and made us very Welcom. We came from thence between 6 and 7 and did not get home till [al]most 11.

29 August. 5 Horses and 2 Carts carried Earth out of Reekyard into Gully. A Shower in the Morning, drying Wind Afterward. Turned all the Barly in Hoglease, no Mowing yesterday nor this day. Yesterday Masons Pointed the Kitchin Windows and Walls up to the East Windo of the Old North Chamber. Today Wetted up the Load of Lime and Earth for Mudwall. J[ohn] Allen's J. Tucker put a Stook to Beer Lane Gate put Bars between Hogleases and

Saltfield and a latch to the Garden door and Mended the South Door of the Wheat Barn. Mr Frampton returned from Farley to Moreton about 3 in the Afternoon.

30 August. 6 Horses ploughed a Bed again in Reekyard and Afterward carried more Gravely Earth with 2 Carts into Gully. After Dinner We Pooked up and fetched in 4 little Loads of Barley. Masons laid another raring on the Mudwalls and some by the End of the Church. J[ohn] A[llen]'s Man, Tucker, mended the Wheel of the Wheelbarrow, the West and North doors of the Wheat barn. A Calm, Cloudy Morning but Clear, Hot and very drying in the Afternoon. We Expected Mr F[rampton] would come hither, but 'twas Thought He went to See the Fleet.

31 August. I, George and 4 Horses with the Wagon fetched a New Dung Cart from Wool which John Allen have made for me and We carried it to Rye Hill and the Wheels are to be bound tomorrow. We called to at Heffelton and We borrowed Father White's Cyder Mill and wring and brought some Malt dust from Beer. 5 Horses with 2 Carts carried dung up out of Barton into Cockroad. It began raining in the Morning and rained very near all day and about 1 'twas very hard. I met my Father in Wool Common going to Woodstreet to Catch out his Weathers *[sic]* which he intend to drive to Wilton or Shrowten fair. We ground up what falling apples We had. Masons Finished the Mudwal by the Church and put some by the Door at House Corner. Slated leak holes in the Brewhouse where it rains down in many places and T[homa]s Fudge laid the Floor in the Bottle house under Stairs.

1 September. Began Ploughing with 2 Suls 7 Horses in Fudge's Close for Turnips. Drisling rain Most of the Fore and some of the Afternoon. We picked up Aples. It broke away dry in the Evening and I sent Most of my hands and Turned the Barly. I rode out into Heath to see for Young Heifers but did not see them. I Rode to Rye hill to see if they had bound the Cart Wheels and they were Finished and We went to Beer and drank together. While I was gone Mr Frampton came and saw the Masons He was in a very

78

good Humor. Mr Jerrit shewed him how it rained down in the Brewhouse and he said they (those that laid up the Stone) must right it. The Masons finished Ceiling and plaistering the 2 Old Chambers and so they have now a done above Stairs and they began plaistering the Passage below.

2 September.   7 Horses Ploughed again till 10 and 1. We Sowed 2 Load of Ashes on the Turnips which look poor as R. Philps's and after dinner Pooked and fetched in 5 Load of Barley from Lower Hoglease. A Calm Morning, the dew laid long. Fine in the Afternoon but about Sunset seemed inclinable for rain. I rode to Beer in the Morning and from thence to Bloxworth with James Bascomb and saw and Bought a Ram of him for a Guinea and put in among the Ews as Soon as I got him home. The Masons finished plaistering the lower Passage and Parlor so they have now very near done within door. They set about the little house next.

3 September [Sunday].   Some Small rain in Morning and several Times in the day. About 11 My Father and Mother came to see us. We Went to Church together. Text Psalms 100.3 and they Sung Psalm 25. This day was Observed throughout the land for a General Thanksgiving To Almighty God, for granting Success to his Majesty's Arms in Taking Louisbourgh from the French. After Prayer The Reverend Mr Fisher Gave my Wife in the Church a Printed Sermon preached by Bishop Beveridge called *The Usefulness and Excellency of Common Prayer.* We Staid and talked about Joseph Alner's Bond which is not good and he is to give another. I wrote a Letter for my Father to Brother Loxly at Martin to advise him that He intends to be at Wilton Fair with some Sheep. My Father and Mother departed about 5. William White of Snelling was at Church and in here too.

4 September.   7 Horses ploughed with 2 Sulls in Fudge's Close and Finished it. We Ground up what Apples We had and pressed it out which by putting Water to it, made an Hogshead. Some drisling Showers and a sort of a Moist Air all the day. No Harvesting. May the Lord in Mercy send us good Weather for the rest. Shoveled up dirt, carried away some Earth before the Kitchin door into the Garden to raise it up higher. After dinner My Wife and son

S[tephen] W[hite] went to George Woodrow's to bespeak some shoes for him. Thomas Fudge and Son Tom, pointed the Windos on outside and stopped up the Scaffold holes.

5 September. 4 Horses dragged and 2 Harrowed Fudge's Close. A Foggy Morning, about Noon broke away good Weather. We turned all the Swaths of Barly Pooked and fetched in 2 little Loads and it rained and beat us off. I sent Robin [Locass] to Heffelton and got some Turnip seed. I rode to Snelling and paid 3 Church Rates and so to Bovington and saw my Father and Mother and Borrowed £10 of my Father. He got in 10 Load of Barly but 'twas pooked before. There was Richard Duel's Father-in-law, Thomas Worsham of Totton. Masons brought down the old Chimney Quine by the Churchyard and dug out Hole for the Necessary House.

6 September. 7 Horses with all the 3 Dung Carts carried Ashes to Fudge's Close, 18 Load from Spear's, [?5] Load from dairyhouse and 7 from Charles Hodges in all 30 Load, 2 Men Filled 'em, George [Sinnick] drove it. 2 Men Stood and Took it out of the Cart and Spurred it without pulling it out into Hills and a Boy stood up in the Cart and put it back to them. This [they] did Over the Plot of 5 Acres Except a little at the South East corner in which I sowed a Sack of Malt dust. When the Ashes was spread, I had it harrowed 1 Time, then I sowed the Turnip Seed and harrowed it again Once More After 'twas sowed. I hope God will Bless my endeavour and send me a good Crop. 2 Men and 7 Women pooked all Hoglease and just as they had done it rained so that they could not do any More. 'Twas near Sunset when the Horses had done Carrying Ashes. Mr F[rampton] was here this Morning and Saw Nanny at Breakfast. He bade about shouting with Mr Plydell till Dinner Time. Several small drisling showers but a Very brisk drying Wind.

7 September. 2 Horses made an End Harrowing in the Turnips and 4 began Plow[ing] in Cockroad for Wheat at the East side of the North Furlong. Began Mowing again in the Afternoon, Turned all the Swaths cut before today and pooked and fetched in 7 Load from Mountfield and 2 from Hoglease which they pooked yesterday and there is more than 1 Load left. The Sun Shone out and a good brisk Wind which made the Corn in excellent Order, but in the Afternoon it grew Cloudy and some drops which though did not hinder carrying yet made it unfit to Pook. The Masons made an [end of] digging out the Necessary House hole and laid a few Brix for the Foundation yesterday and today rise the Wall of the Valt 5 Feet put the Drawers and Table into Kitchin Dined, Supped and went to prayers in it.

8 September. Small rain. 6 Horses, 2 Sulls ploughed all day in Cockroad. Several Showers of Rain but not much here yet, a Great deal in some places. I rode to Bovington and saw my Father and Mother and son J[oseph] K[ingston]. Father was Ill of a Cold of which he Ailed the day before but now Seemed better. I rode to Heffelton and saw Mother White, Sister Patty and Brothers S[tephen] and Roger and from thence I rode further on in the Heath and found out 7 Heifers I had out there and brought 'em home into Ground. Did Nothing about the Corn Except Mowers who kept on Cutting. My Father sent me a Ram with 80 Wether Sheep to go to Wilton Fair next Tuesday. Masons ris[e] the little house 3 courses above the Hole behind.

9 September. Philip Burden set out with Father's Sheep by 5 and I went with him to Beer Cross, he is to Lodge at Brother Loxly's at Martin Anon. 6 Horses, 2 Suls Plowed till between 11 and 12. Women picked up Aples in the Morning, Pooked and Turned Barley. We brought in the rest of Hoglease 2 Load More and 7 Load from Mountfield haven't but about a Load out of what was cut before Thursday. Mr F[rampton] was here in the Evening and saw the Masons who set in the Door Durns and got up the Wall 16 or 17 Courses above the Threshold.

10 September [Sunday]. I and Nanny Went to Beer Meeting there was read Psalm 135 and 2 Corinthians 6 and they sung Psalm 135. Text 2 Corinthians. We Went to Church in Afternoon. Text 1 Corinthians 3.3,4 and they Sung Psalm 105. Mr Fisher Saw us as We came from Meeting and He preach much against seperating from the Church Established. My Father came while We Was at Church and Staid a little While. I Went with him as far as Robert

Philps's and shewed him my Orchards and Turnips. Cousin R. Porter was just in here after prayer. The Dairyman, his Wife and Children was here in the Evening. A Blessed Day for the Corn.

11 September.   5 Horses with 2 Carts carried Dung from home into Cockr[oad] and 4 ploughed there with 1 Sull. 3 Women picked up aples and Turned Barley, did not Carry or Pook any. I Set out between 1 and 2 For Martin and so to Wilton Fair. I got to Martin between 5 and 6. I Pact my Horse at Brother Loxly's and supped there with Brother Michael Saunders and his Wife. I went before Supper and Saw Brother and Sister Brown. They have been both back again from Handley some Time whither they was carried with an Order and They are to bid quiet Till Michaelmas till they can get another place. P. Burden Set out in the Morning with the Sheep from Martin for Wilton. Blessed fine Wether.

12 September.   Brother Loxly and I rose Early and about 3 Set Out for Wilton fair and Got there and Met with P. B[urden] driving the sheep to the Prize which we penned before Sunrise and I Sold 'em by 7 Clock or Soon After for just 16s. Each which 80 came to £64. I Sold 'em to Mr Oliver Jeffery of Chichester. He paid me 35 of 36s. pieces and 1 Guinea out of which I paid him 1s. We came out of Town about Noon. Called at William Lanham's and dined with him and got back to Martin about 6. I and Brother M[ichael] S[aunders] went to Brother Brown's and spent a little Time with them, then Brother Brown went with us and supped together at Brother M[ichael] S[aunders]'s with Brother and Sister Loxly and about 9 returned with them and lodged at Their House and as I went back staid in a little again at Brother Brown's while he Clove and put up a Loaf of Sugar for me to Bring for Father White at Heffelton.

[Interlined between 12 and 13 September] Horses Dung carted and plowed as Yesterday and After dinner the folk pooked some barly in Mountfield but carried None. A blessed fine day.

13 September.   9 Horses ploughed and Dung carted till between 10 and 11 and then fetched in 8 Load of Barley, 1 from Mountfield, 5 from Pit Close, which was all that grew there, and 2 Load out of Close by Hodges's. I rose about 4 and about Sunrise

Took my Leave of Brother and Sister Loxly and got home about 11 and the Folk was just got to Pooking and Carrying. Just after Mr F[rampton] came and saw the Masons who Finished the Walls of the Necessary house and did something to the Wall at East of it. Fetched in 1 Load of Barly from Mountfield, 5 all Pit close and 2 From Lower 6 Acres.

14 September. Began carrying Barly in Morning, the last of Mountfield 5 Load and all the rest of Lower 6 Acres g[round] and from Spear's 2 Load, We pooked and was brisk with it. 'Twas Cloudy Wether for the Most part dry and Good. Masons about the Wall. Mr Manuel called to Tell he drove out some Sheep from my Turnips at R. Philps's. W. Curtis began Breeching my son James Lyne.

15 September. Nanny rode to Beer Early before Breakfast. A Blessed day again for the Harvest. We Made an End Pooking all our Corn and got in All the Barly some of which I hope will yield very Well but it is put in very Close together, for I have but 58 Load out of 54½ Acres. From Hoglease 15, Mountfield 20, from Spear's 6, Pit Close 5, Lower 6 Acres 11 and Turnip Close 1. Women Raked Rushes and picked up Apples. Afterward I and Tom Philps Surveyed Pit Close and all the Plots by Hodges's. Father and Mother White came about 5. They made an End Harvest yesterday, they departed about 8. George [Sinnick] took Ill this Evening.

16 September. 4 Horses Ploughed and 5 Carried away Brix before the Door and Batts into Mead to the Fords and Rubble up into Cockroad. Mr F[rampton] came round here a Shooting. I and Tom Philps measured the rest of the Ground the Mowers cut which I believe won't answer 5 Acres to what they Account 6, so that was I to give them 1s. for the Lug Acre It would not cost me as much as it do now at 10d. My Mowers are displeased because I've measured it. Blessed fine Wether. Masons finished covering the Necessary.

17 September [Sunday]. I thought to go to Beer Church in Forenoon but rained so I did not. About 2 Sister-in-law Mary Saunders of Harbridge and her Son William came to see us but was very Wet. They went to Church with me. Text Luke 13. 24

and they Sung Psalm 13. There was but 22 folk at Church, Children and all. Poor Mr Henry Jacob of S*[space]* had a great loss by Fire last Friday Night, his Barn with the Corn and Out house with Timber was burnt but not his house. A dry Evening.

18 September. 4 Horses Ploughed and 5 Carried more Ruble up into Cockroad. Charles Hodges had promised a sack of our pears and carried 'em up early this Morning to Woodbury hill, but the folk that bespoke 'em would not have 'em and he sold 'em to Beck Reason for 3s. and 3d. I reckoned with my Work folk and paid 'em as the ground laid for before. I rode to Beer and bought some things and after I came home I rode to fair with Mary Saunders and her son William and there We met with my Father and Son, J[oseph] K[ingston]. Father and I Sold our Wool to Mr Taunton for 22s. a wait only Father gives 2 fleeces into the Bargain. J[oseph] K[ingston] came hither with Me. Nephew W. Saunders and his Mother Went with Father to Bovington.

19 September. 6 Horses fetched a Load 800 Brix from Oakers Wood and afterward carried Earth up into Cockroad and drew away Old Brix to the Pile and in the Evening went and helped Farmer Alner carry som Yeegrass Hay. About 11 Mr F[rampton] was here and between 11 and 12 Brother S[tephen] W[hite] came and rode to Fair with I and Nanny. I carried some Pears for the Man at Hove and He would not have them. My Father was not at fair, there was Mother White, 2 Brothers, Sister Patty, Brother and Sister Florence and all there Children. Nanny came hither with us. William Hurl of Verwood and his 2 Daughters and Molly's Husband was there and I believe some of 'em are gone to Bovington. I saw Cousin Joseph Effemay and he said he would come with me but I lost him and he came not. Fine Wether till the Evening and then a little rain.

20 September. 4 Horses ploughed in the Morning and 5 carried Earth up into Cockroad about 11 Shut off Cart and fetched in the Oats by Hodges and had 1 Load in 5 Acres. The folk went to Fair Afterward I, Nanny, J[oseph] K[ingston] and Nancy Florence rode to Woodbury hill fair. Saw and Dined with Father White and Met with Father and drank with him and Mr Bestland and

J[oseph] K[ingston] rode with Father to Bovington. Brother S[tephen] W[hite] did not go home last Night. We put the Bed in which I and Nanny lies into the biggest of the New Chambers and intend to lie there for the futre. Clouds about but dry.

21 September.   9 Horses carried Earth and finished the little furlong at East in Cockroad and fetched in a Load Fagots for the Dairyman. I rode to Bovington and Carried home my Mother's Great Kettle. After I came back I and Nanny rode to Woodbury hill Fair again. I had up Dragon the Poor Horse but did not sell him. Very Cold and some Showers. There was Mother White, Brother Roger and Uncle John, Brother and Sister Florence and there Daughter Betty. I bought 5 Locks etc., Nanny a Tin Oven and Changed a Pewter Dish etc. Mr John Taunton of Wrackleford and his Man came yesterday Morning and packed my Wool which was 4 Weight. 10 lbs and Lams Wool, 1 Weight 2 lbs. at £1 1s. a Weight. so I had £5 18s. 6d. for it and 6d. which I gave the Man. Mr Aldridge of Moreton came to see my Vetches just after they was gone and staid with me a little.

22 September.   4 Horses ploughed but did not quite finished the East Furlong but ploughed a little in drot and 5 Horses carried dung with 2 Carts from out of Barton. I, George [Sinnick] and J[ack] Hodges Winnowed Vetches Since Dinner. Nanny and Nancy Florence have been again at Fair this Afternoon. A Cold Morn, Wind at North, a Shower between 5 and 6. The Masons finished the Wall between the House and Necessary and Tipped part of the Mudwall Over the Door. Yesterday J[ohn] Allen's Men began putting the seat in the Necessary House, are to get out Door Durns to day at Bovington to put by the Church Hatch again the little house which are to fetch tomorrow.

23 September.   4 Horses Ploughed in the Drot and I and George [Sinnick] with the Wagon and 5 [horses] fetched the New 2 Hatch Wear from Bovington to set in the New Main at Furze Goalds and brought some pieces to make door Durns which J. Bowden made After We got here and Simon Hallett finished the Seats of the Privy house. The Masons rited up the East Wall of the Garden and laid a Raring of Mud to the Cart house by the North door. Since dinner Sent George [Sinnick] and Robin [Locass] to Moreton with 7 Bushels of Vetches for Mr Aldridge. Dry Weather.

24 September [Sunday]. I, Nanny, Nancy Florence, Sons, S[tephen] W[hite] and J[ames] L[yne], went to Church. Text Luke 13. 24 and they Sung Psalm 66. 'Twas a blessed fine day. My Father came just as we came from Church to let me know He would not go to Shrowton fair tomorrow. Father departed abut 5. William Bearns of Wool was here and fetched some Pears.

25 September. 4 Horses made an End plowing the drot and began the Great South furlong and 4 Horses with 2 Carts carried dung from the Mixon in Cockroad. I rode to Shrowton Fair, saw Father White and Brother Roger and Uncle James Dewel. 'Twas allowed that Sheep did not sell so well there as at Wilton. My Father Sold his 40 last Wethers at Home last Tuesday or Wednesday to Farmer Lovelass of Melcomb for 14s. and 6d. a piece and was to be paid for 'em here this day and as he did not Go I was to Take the Money but I did not see the Man there. Nanny rode to Heffelton to See Sister and Brother Florence but they was not there. I came home with T. Barratt of Wooll so far as Beer. Fine Wether.

26 September. 4 Horses Ploughed and 4 with 2 Carts carried dung as yesterday. I put my 13 young beasts out into Heath again and they are to lie in Ground at Night. I and Butcher Bascom of Beer drove Each a Cow (most Fat) to Bovington to keepring. Father rode to Heffelton to Dinner and Meet Brother and Sister Florence about dividing among 'em what Brother N[athaniel] L[angford] left. We Staid and dined with Mother and called at Brockhole and there the Butcher had a Calf about 5 or 6 Week Old for which he gave a Guinea. Picked in the Aples here at home which was 6 Sacks or more. Masons finished the Walls by the Churchyard Floored the Necessary house and pointed Some Windows. The Carpenters hung a door to it yesterday to that 'tis now compleat. A little rain in the Night but a good dry day.

27 September. 3 Horses with the Wagon fetched 25 pair of Small Poles from C[harles] Hodges and Afterward 6 Horses and 2 Carts to Waymending, they went about 9 and gave Off about 1 Such a

Hand they make at it. Brother Roger White brought us *[space]* lb of Bull Beef. John Allen Came and Said he had been at Moreton and Mr F[rampton] would not pay him his Bill till he had Spoke with me. After Dinner we gathered in almost 5 Sacks of Aples in Cicilas Orchard, the While Mr F[rampton] came and asked the Mason for us but He did not ride round that way so I did not see him. Jarrard the Mason began the Underground gutter from the Brewhouse Sink and done it round almost to the Kitchin door, 'tis to be carried down into the Garden. A Thick Fog in Morning but very fair Afterward. We got in above 5 Sack of Aples last Night in our home Orchard and above 3 Monday Night out of the dairy Orchard.

28 September.   4 Horses ploughed in Cockroad and 3 with 1 Cart Waymended and 2 rested. About 2 Nanny rode to Bovington to See my Father and Mother. I rode to Beer in Morn for things. Went round Saw Sheep, Plowfolk, Turves and Waymenders who differed at Longbridge. After dinner squeezed out Water Cider. Fine Wether but about 2 a Shower. N. G[arrard] the Mason finished the underground Gutter.

29 September.   8 Horses with 2 Wagons carried 6 Load of Turves to Beer but 'twas Late (7 Clock) before they got home. I rode to Bovington and Saw my Father and Mother and Directed Butter Tubs for him to Ringwood. Robin Locass's Time being out I paid him and he is gone to live with Farmer Cooper a[t] West Creech. After Noon I and Old T. Hunt beat up the Ford at Gully and let the Water down through into Pin Mead. Farmer Nathaniel Swetman of Buckland Lodged in Farmer Willshear's Hoglease, 671 Sheep for Way hill fair. N. G[arrard] the Mason Sorted out the Old Brix by the Church yard and Groved out the Parlor Windo for a Casement but the Blacksmith did not bring it. Frost in morn but fine Wether Afterward. Sister Patty came about Sunset to go with I and Nanny to Martin tomorrow, Brother Roger came with her but he did not Stay long.

30 September.   6 Horses fetched in Turves, a Load for the Dairyman and 1 Load for ourselves, out of the South heath. About 10 I, Nanny and Sister Patty set out for Martin and got there at 4. We Staid near an hour in Blandford, saw Uncle James Duell there. He Sold a Load of Wheat There to Talbott of Wimborn for 10 guineas. We Saw a Vast Number of Dorsetshier sheep going up to Wayhill fair and saw Mr Bridle of Milborne, Cousin William Tilly of Elsington and Farmer Antrem of Lower Woodsford. 'Twas Cloudy, Mild Wether and the Roads very Good.

1 October [Sunday].   We spent the Evening at Brother M[ichael] S[aunders]'s with W[illiam] Tilly. I and Nanny Laid at Brother H[enry] L[oxly]'s and Sister Patty at Brother Brown's. I and Brother M[ichael] S[aunders] and Nephew J. Loxly went to Prayers in the Morning and after Dinner I, John and Elizabeth Loxly rode to Fordingbridge Church. Text Ecclesiastes 1 *[space]* and they Sung Magnificate and Psalms 1 and 40. They have a Bassoon and after Service was Over they Sung several Anthems and Played on the Bassoon, a German flute and the Violin which was Noble Musick together. Sister Patty was going with but William Tilly got her away to Salisbury with him. Text at Martin Church Luke 22. 1, here at Piddle Proverbs 28.13.

2 October.   6 Horses fetched 2 Load of Turves, 1 for the dairyman And 1 for ourselves and 3 Horses with 2 Carts carried abroad dung in Cockroad. I, Nanny and Sister Patty was going to Set out from Martin but Sister's Horse was ill, so they stayed and I rode to Salisbury to See if Mr Brander was there but he was at Nea. I was at his Mother-in-law's, Mrs Shaw's, in the Close. She made me very welcom, Treated me with Wine as did Mr Commissioner Hooper who was going from Boveridge to London. I heard on the road that he was before me and I found he Quartered at the Antelope and he was just going to Take Chaise when I saw him. He was Exceeding affable to me. I went to Evening Prayer at the Cathedrall and was up with the Organist in the Organ Loft while He played. Many of the Dorset Sheep was about this City to Night. John Antel of Portsham, a lad of about 19, Offered his Service to me, he is going up with Sheep and will call as he goes back. Fine Wether.

3 October.   6 Horses carried 4 Load of Turves for Charles Hodges and 3 with 2 Carts carried abroad more dung. I, Nanny and Sister

Patty Left Martin and set out for Ringwood. We called at Harbridge and Staid about of an hour with M[ary] Saunders, not 1 Child was at home but she went down with us to Kent* and We saw 3 of 'em there and drank a dram at Farmer Biles's. Then we called and drank some Ale at Farmer Hunt's at Ibsley and from thence rode and dined with him at his Son-in-law's Farmer William Mist's at Moyls court and saw the Garden and Wilderness and drank Tea, then rode through Powlner to Ringwood and then to Cousin Joseph Warn's at Kingston. There We Supped with Young Cousin James and Elizabeth Warne and went back to Ringwood with them leaving our horses there. Blessed Wether Continues.

4 October. 6 Horses carried in 4 Load Turvs for Simon Philps, 2 out of South and 2 Out of the North Heath and the dairyman had 2 Horses and his Own Mare with a Cart carried his Tub Butter to Wareham. I, Nanny and Sister Patty Laid all at Cousin James Warn's at Ringwood and We went and Breakfasted with Cousin Henry and Sarah Jessop and Afterward went up to Cousin Thomas Russell's. Cousin John R[ussell] was gone to Salisbury to the Festival of St Cicillia so We had none of his Company. We 3, Cousin T. R[ussell], H. Jessop and his Wife, all dined together at Cousin James Warne's and then Cousin T. R[ussell] and H. J[essop] went to Kingston with me and fetched our Horses, the While Nanny and Sister Patty went with Cousin Elizabeth Warne and saw Mr Willis's Wilderness and Hermit's house and Mr Pitt's Organ and heard him Play on it. About 3 We took our leave of them and made no Stay on the Road and got home between 8 and 9. Fine dry Wether Continues.

5 October. Dairyman with 4 Horses Ploughed in Cockroad and George [Sinnick] with 4 [horses] ad 2 Carts Waymended. I rode to Bovington and had the Sucking Colt thither and left it there to be Weaned. Father was at Bindon so I did not see him. I Picked in Some Aples After I got home. Fine Mild Weather Continues.

6 October. Dairyman with 4 Horses and 1 Wagon carried Turves for Robert Philps and was very much Standed in the Gutter and Moild about with them. George [Sinnick] has Battered

* An isolated farm in Harbridge parish

89

about the Horses so that but few of them will stand to draw when but very little stopped. George with 3 [horses] and 1 Cart Way Mended. I drew out Pin Mead ditch Edged up the dirt at Barton Gate and dug out the South rout and they filled it up with Gravel. We put Stones at the End of Gully Bridge to keep up the Gravel and turned all the Water under it and made good the Bank by. I and Nanny rode to Mrs Samways's and saw 4½ hogsheads which I am to have for 16s.

7 *October.* All 9 Horses with both Wagons fetched in 6 Load Turves for William Curtis and a Load of Furze for ourselves, but they was late about it. It seemed inclined for rain in the Morning so I had some Turves which I had cut last Saturday and Monday in Farm Heath Pooked up, but the rain Wore off and dry Wether continued. About 10 Brother Florence and Cousin Nathaniel [?Florence] came and carried away Nancy and I rode with them to Wareham Market. I Sold a Load of Stained Barly there to Mr Bestland for 21s. a Quarter and am to have more if he can get it at Poole and I am to have a Load of Seed Wheat of him for £12. I am to Carry the Barly and fetch the Wheat.

8 *October [Sunday].* Some Showers in the Night and Morning. I rode to Bovington and dined there with my Mother and son J[oseph] K[ingston]. My Father was gone to dine with Mr Frampton. I rode to Wool Church. Text 2 Timothy 3 and we Sung there Psalms 66, 8, 149/132^a, 26^a, 16^a and Communion responsals. I called at Bovington as I came back but Father was not returned. Text here at home Deuteronomy *[space]* and they Sung Psalm 33. Stormy in the Evening and Wind Very high.

9 *October.* 4 Horses ploughed in Cockroad and 4 with 2 Carts Carried ruble and Stones and Gravel and made a Rolling Bay at the Corner of Lillington's Mead. After dinner got in More Aples. John Antel of Posham or Portesham a Lad of about 19 called and laid here last Night as he came from Wayhill where the Sheep Farmers had the best fair 'twas said that was ever known and this Morning I bargained with him for 4 Guineas. He is to Come next Monday and serve me till next Old Michaelmas day. The Wind was Very high all last Night and some rain.

10 October. 4 Horses ploughed and 5 with 2 Carts carried abroad dung in Cockroad. I rode to and from Ower fair with the 2 Butcher Bascombs. I Saw my Father and Mr Cockram there Who have Bought my Father's Old Fat Cows and mine and he will send me 40 Ews next Saturday or Monday. 2 or 3 Small Showers but else a good day for the Fair. A Great many Pigs there but very dear, no Cows and but few if any horses. Tom Hodges came to live here in R[obin] Locass's stead. Got home between 7 and 8. Cousin Jonathan Fudge and his Wife and Jenny Willshear here.

11 October. I and George [Sinnick] began Sowing Vetches in Cockroad East Furlong, about 7 Bushels, 6 Horses dragged 'em in, then 2 Harowed 'em down and 4 ploughed a little. I, Simon Philps and Jack Hodges Winnowed 11 Quarters Barley. We did not begin it till 11 or 12 Clock and was late about it. The Dairyman, C[harles] Hodges, Clark and Joseph Boyt made Cider here with Father White's Mill and Wring, about 2 Hogsheads of the best besides Water Cyder. 'Twas blessed fine Wether, very pleasent.

12 October. 7 Horses carried 10 Quarters Barly to Wareham for Mr Bestland but 'twas Shot at the Great Malthouse. I Was to have had a Load of seed Wheat of him but 20 Bushels which was there a Saturday was Sold and carried away and the rest was not brought by reason (I think) 'twas such a Very Wet Morning. The Wind at North East and Very Cold. I called at Heffelton as I came back, dined and drank Tea and staid there till 7. I had 4 Men a Cyder making, made a Hogshead. A dry Afternoon.

13 October. 4 Horses ploughed with 1 Sull. 2 made an end dressing in the Vetches and 3 rested. I, William Curtis, Charles Hodges and Betty Kite and Thomas Hunt, John Thomas Hodges help a little about making up the rest of our Cider and we Finished and Geffer Rawls's too. We wrung out 5 half Hogshead of the best and 1 of Small with what was made Yesterday and before We have 4 Hogsheads and ¼ of the Best but my Father is to have ½ a Hogshead of it. The Water Cyder they shall Wring out tomorrow. We are sorrilly put to it for Barrells. A fine pleasent day but the Wind at North East continues cold.

14 October. 6 Horses ploughed with 2 Sulls in Cockroad North Furlong turning back what they Ploughed at first with 2 Coulters. Winnowed a little wheat in the Morning. I with 3 Horses and the Cart carried ½ a Hogshead of best Cider and ½ a Hogshead of Small and brought back 2 Emty Barrels and 4 Posts. Just After I got here came Cousin Joseph Effemay of Lye by Wimborn and his Eldest son, Joseph. They came from Dorchester. I was about to have rode to Wareham Market but was so late from Bovington that I did not. The Wind very high and some rain. Wrung Out the Water Cyder of which there is about *[space]*.

15 October [Sunday]. Rainish Air in the morning. Cousin Joseph Effemay and his son departed about 9 for Bovington but did intend to Go home by Night. Nanny rode to Beer to go to Meeting there for which I was Very Sorry to find her so bent, but Blessed be God There was no Meeting this Afternoon, had there been any this would have been the 1st Time we should have been I at Church and She at Meeting at 1 and the Same Time so We went to Church together. Text Mark 3. 3 and Mr Fisher made an Excelent sermon. They Sung Psalm 147. My Father and son J[ames] L[yne] came just after we got to Church and departed about Sunset.

16 October. 4 Horses ploughed and 5 carried dung abroad in Cockroad. I rod[e] to Beer in the Morning and fetched some Corks etc. Mr Talbot of Shitterton came and looked on some ditches in Derrick's Mead which want to be Took up. I expected Mr F[rampton] here but he did not come. Joseph Alner fetched our Cyder Mill and Wring to make up their Cyder. Since Dinner William Welch and his Son helped burn and Strap my 15 Calves and Bleeded 'em all. We 1st bleed 'em in the Neck then burn a hole with a red hot Iron in the dew flap and put in a Tan Leather Strap and Tye both ends together* so they are done, this is to prevent the Kill Calf in which We lost 1 yesterday and lost 1 before. John Antle of Possham with whom I bargained last Monday came

* This technique, known as a seton, was used to keep a wound open so as to allow puss or other harmful discharges to pass out along the strap. The disease may have been due to a parasite as it seems to affect cattle only in certain areas and not others.

this Evening between 7 and 8. Dry Wind. James Lyne Warne Born October 16th 1755 at Woodstreet.

17 October. 5 or 4 Horses with 2 Carts carried Gravel, Clots and dirt into Lillington's mead to rite the Bank at head of it. 4 Horses with the Wagon carried home the Cyder Mill and Wring to Heffelton and almost ½ Hogshead of best Cyder and brought back some heath into Mead to make fords. Fine dry Weather continues the wind at North East, Brisk and Cold. Farmer Joseph Alner here in the Evening to Borrow Bags till near 9. Talked much Nonsence Stuff. *[inserted in red ink]* Christopher Warne died at Bovington 17 [October], 1755, aged 64.

18 October. 4 Horses ploughed in North Furlong Cockroad and 4 carried more Gravel and Clots with 2 Carts into Mead. I Work ther too most of the day. Wether Continues [fine] but grow Cloudy. Simon Hallett and Andrew *[space]* yesterday and this day put up a Chimny piece and Shelf, A Calvs Rack at Lower end of Garden and the Jack to roast with and I cleaned it in the Evening. Monday was my 3rd Son James Lyne's birthday he being then 3 year Old and this is my Eldest son's. He is 6 year Old and they are both very healthy Children, blessed be God. *[inserted in red ink]* Joseph Kingston Warne Born 18, 1752 at Bovington.

19 October. 4 Horses ploughed and 5 carried abroad the last of the Mixon with 2 Carts. I Rode to Bovington and Carried my Mother Some Silk and Turnips and drank Tea with her, then I rode down to Bindon and up round to Woodstreet and Back through Wool and brought home some Iron Traices I had there, and T[homas] Barratt and his Wife Offered to lend me Money and I don't know whether about Christmas I might not have £40 to £50 of 'em. I Staid a little at Bovington as I came back, my Mother Gave me a Rabit. Father was very buisey getting in Broad Clover. He has a bundance of it and 'tis very Well seeded and it have been blessed Wether for it Cloudy dry Wether. Mr Aldridge came to see what work J[ohn] Allen had done here and Afterward Mr F[rampton] was here and left word I should go and speak with him tomorrow. John Harding the Cutter came and Cut my Colt and 2 Lambs. Mr Cockram sent me 40 pretty Ews. Joseph Allner brought home Sacks.

20 October.   4 Horses ploughed and made an End of the North Furlong turning it back and 5 carried dung with 2 Carts from home and carried the Fold out of Cockroad Over to the 6 Acres at Buckshill and penned the Sheep there at Night. I rode to Moreton and spoke with Mr F[rampton] about a Wear Farmer Alner is going to put at Upper end of Gully. In the Evening I and Nanny went and spent the Evening with Farmer Alner and his Son and Daughter Shepherd and 5 of his Grand Daughters. Came home about 8.

21 October.   4 Horses ploughed and 5 carried dung from home with 2 Carts. I and Nanny rode to Wareham Market. I sold a Load of Wheat there for £10 5s. to be carried any day next Week and then I shall bring back a Load I am to have of him *[Mr Bestland, see 12 October]* for Seed and gives him £12 for it, which is a Vast difference. We saw my Father and Mother White there. Nanny and her Mother and Sister Florence went and drank Tea at Mr Reader's and invited 'em to come and see us next Wednesday. Cold Wind at East, continues dry. We got home about 7.

22 October [Sunday].   I and Nanny rode to Wool Church in Forenoon. Text Luke 16.31, sung Psalms 23, 108, 26ª. My Father and Son J[oseph] K[ingston] was there. My Father, I and about 20 more Staid and received the B[lessed] Sacrament, J. Chaffey too the 1st Time I ever Saw him. We rode with Father to Bovington and dined there with him and Mother and Galton and his Wife. We came away between 2 and 3 and I Went to Church here. Text Mark 3.5 and they sung Psalm 25. Brother S[tephen] W[hite] called here as He went home from Beer and staid till 10. I had another Calf died in the Kill Calf. Close Cloudy wether.

23 October.   The 9 Horses with both wagons fetched in 4000 of Turves out of South Heath for our Selves. I and C[harles] H[odges] skinned the Calf after dinner. About 3 Miss Becky and Miss Jenny Ekins's came to see us and staid and drank Tea with

94

Nanny. It seemed likely to rain in the Morning, but it broke a Way, the Wind Continues at East.

24 October. 5 Horses and 2 Carts made an end dunging Cockroad and 4 ploughed and [al]most finished plowing it. The Carts Afterward carried more Clots and dirt to the Bank at the head of Lillington's Mead. I rode to Beer in the morning and carried my Mother some Turnips and brought home some Lime to Lime my seed Wheat for I think to begin sowing tomorrow. Mr Aldridge came in while I was There and Staid a little. It rained in the Night and kind of Moist drisling air all the day. Wind at North. Mr Frampton Appointed to come yesterday or today but hasn't been here. Mr Spond, Mr Humphrey's Clark, was here and held Court* in the Forenoon.

25 October. 3 Horses rested and 6 dragged before We sowed Once in a place and Twice after. We sowed 8 Bushels and that did *[space]* ridges. Mr Frampton and Mr Milborn came round here in the Forenoon and looked on the Place where Farmer Alner is to set his Wear and down on the New Work in Lillington's mead. And in the morning here was Old William Talbot of Shitterton and had a Letter of me to carry to Mr Derrick in the Isle of Wight offering to Take his Living here. The dairyman and T[homas] Hunt Told me yesterday of a Plot of ground there is in Alner's Coppice Close belonging to Fords. I wrote to Derrick thereof and Also Told Mr Frampton of it who went and saw it and will lay claim to it, just as he went away came Sister Florence, Mr Reader, his Wife and Daughter and a little After Brother Florence, Brother Roger and Mother White and while We was at Dinner Father White and before we had done, My Own Father and Mother. All Wareham folk departed between 4 and 5, our Fathers, Mothers and Brother R[oger] a little after 5. 'Twas a fine pleasant day.

26 October. 4 Horses carried dung with 2 Carts from home into Hilliar and 5 carried a Load of Wheat to Wareham and brought back a Load for seed. I and Jack Philps drove 11 of my Calves over to Bindon, for Father said when here yesterday If I would keep

* The manor court: possibly Briantspuddle Manor, which belonged to Mr Humphrey of Wimborne (*v.* J. Brocklebank, *Affpuddle* (1968), 39).

some 2 Yeared Beasts for him here he would keep my Calves at Woodstreet where they never dies of the Kill Calf. I measured some Ground there which George Read routed in Old Harris's Lott 32 Lugs and there's but very little more to rout of the whole would not [be] above 2 or 3 Lug. It rained pretty well all last night so that the Waters are rise[n] high in the rivers but 'twas a fine dry Sun Shine day. Father came into Wood[street] to us and I rode with him to Bovington and my Mother gave me some Bull Beef and Tripe.

27 October.   I sowed 12 Bushels Wheat which almost finished that South Furlong. 5 Horses dragged before and After 'twas Sowed, then 3 ploughed in the North Furlong and I Struck up Furrows with the other 2 and 4 Horses Harrowed and finished the South Furlong, Except the Headland and part of a ridge by the North F[urlong]. 'Twas a frost in the morning I saw Ice in the Pond by the Pump, the 1st I've seen this Season. Wind at South but a dry day, we put in the 4 Stones at the New Bay at the Corner of Lillington's Mead and Took the Water into Alderham.

28 October.   I sowed a little more than 4 Bushels which Finished the Drot and finished dressing it and dragged Some of the North Furlong. It began to rain about 8 and have rained pretty well ever since we shut off between 11 and 12. Cut down Barley Mow and Wheat Mow. I and Nanny Testered up our bed and put on the Vallens. I and George [Sinnick] Tailed the Eel net caught but about 6 or 7 Eels, some of 'em pretty Large and a pretty Trout in Farm Mead North Wear, 'Tis a good likely Night for them to run. Between 6 and 7 Our Child Robert Duel can go very pretty by the Wall and Chairs but not to run off a Cross the house.

29 October [Sunday].   I and Nanny was about to go to Beer Meeting but 'twas Stormy so We went not and as it happened there was no Meeting. We went to Church in Afternoon. Text Mark 3. 5 and they sung Psalms 105. Cousin William Fry was at Church and in here Afterward. My Old Servant Daniel Cob was also at Church and in Stable but did not Speak to me. See August 4th.

30 October. 3 Horses carried 2 Load of Dung up Into Cockroad and Afterward Restered in Hilliar's Plott and 6 made an end dragging the North Furlong in Cockroad and harrowed some and ploughed some in the headland. I and Nanny rode to Puddletown fair on the Dairyman's mare and he rode my mare. I let my man John Antle ride the Old Mare yesterday to Possham and he did not come home till this Evening I saw him at Puddletown. He Staid there to See the fair but I was not very Well pleased with him. I caught 8 Eels last Night and some very large, more than a lb. Each. We got home from fair between 6 and 7. A fine day but rain again in the Evening.

31 October. 3 Horses Restered again and 6 Carried dung from Home into Hilliar's plott and brought back Earth into the New garden, so we had 3 Carts, 3 Under Horses and 3 Traice Horses, 1 Car afilling at Each place and 1 always going Loaded. They carried *[space]* Load of dung and brought back *[space]* Load of Earth. Nanny was Took Ill about 9 or 10, so we sent for Kate Curtis who came and bled her and she seems somewhat better this Evening. It rained very hard before we got up and several showers in the day. My man John Antle pissed a bed Saturday night and Sunday Morning before He went home He made up the Beds in the Garrett but the Maidens had a Suspicion and found it out yesterday before he came home. Caught but 1 Eel last Night.

1 November. We carried dung with 7 Horses and 3 Carts, 3 Load from Dairy house and Afterward from the Mixon in Long Craft into the 6 Acres. We Carried 30 Load in all. I, C[harles] H[odges], T[homas] H[unt] and John Antle filled, Jack Philps's drove and George [Sinnick] pulled it out. Mr Pickard of Bloxworth and his Brother Mr John Trenchard, rode by going to Warmell and put into Cicilas Cart house for Shelter out of a Shower. It began to rain again about 2 and has rained fast ever Since. The Water rise in the River. I did not put in my Eel net last night but have this Evening, at the Lower Great Wear, the 1st Time We ever put in there.

2 November. Caught no Eels. 4 Horses made an End restering Hilliar's and 5 carried dung in Long Craft with 2 Carts. C[harles] H[odges] and J[ohn] A[ntle] filled and they carried 14 Load and

1 from Spear's. I Levelled the Hedland in Spear's and rode to Bovington and Carried my Father and Mother Some Turnips and dined with them and in Woolbridge heath found my Heifers and brought 'em home into Farm Mead and put in the Eel net there at the South Wear. Tuesday Night I Finish a Copy of a Poem on Dorchester. Much rain from 2 yesterday to this Morning, but dry for the Most Part Most of this forenoon, but about 3 it began to rain again.

3 November. Caught 12 or 13 Eels but 2 or 3 was Very small. All the 9 Horses and 3 Carts carried abroad 26 Load the rest of the Mixon in Long Craft, Except about 2 Load. The Ground is so Very Wet and the Wheels cuts in so that they are often standed and George [Sinnick] broke another Dung Cart draught. I Told 'em after dinner He should go to Threishing or Take his Wages and go a Long but he said He should not go to Threishing and I Talked with him afterward in Stable and he said he would go to days Work. I must part with him for 'twill not do to have so many things broke. They carried *[space]* Load in long Craft and 2 from Spear's. Several showers this Afternoon. I made an Eel net frame and put it in again at Farm Mead South Wear. Robert Fudge and Tim Hooper Tipped up the Mud Wall and rared the piece at Barton Gate and began a Sink by Brew House door.

4 November. Caught but 2 Small Eels and about 38 Minice. C[harles] H[odges] with 3 Horses harrowed down the Resters in Hilliar's Plott and 6 with 2 Carts carried Brix and Mortar to Dairy house and about 9 or 10 Load of Dung out of Cicilas Barton. George [Sinnick] Filled it and I made Thomas Hodges drive it. I cut Gutters in the Stones at Door. R. Fudge put in the Iron Grate by the Brew house door and ris up the Garden Wall a little at Dairyhouse. Several showers again this day. I rode to Beer Since dinner and bought some Things and from thence to Wool and got a Draught for the Dung Cart and brought it to R. Philps's.

5 November [Sunday]. Nanny rode and I went to Beer Meeting. Text Psalms 111. 10. We all went to Church in Afternoon. Text Psalms 118 *[space]* and they Sung Psalm 16th. Several Showers again this day.

98

6 November.   3 Horses ploughed in Buckshill 6 Acres and 6 with 2 Carts Carried 10 Load of dung into the Same Ground, 1 Load from Spear's and 9 from Cicilas. Mr Plydall with 2 Men and 2 Boys came a Rabet Hunting about there, Some he shot, some the Dogs Caught and some we Ferretted out, in all 14. Nanny rode to Heffelton and I did not come in from Ferretting till about Sunset. Brother M[ichael] Saunders came about 4 and left his horse and was just going to Bovington as I came in, so Staid about ½ an hour with me and just after Brother S[tephen] W[hite] and Told me that Nanny did bide at Heffelton all Night. I was about to have rod[e] thither and came home with her but on hearing she did stay I lent my mare to Brother M[ichael] and he rode her to Bovington. 'Twas a smart frost in the Morning but soon went off with a little rain was Afterward tollerable and dry till about 6 and now it rains again apace.

7 November.   All the 9 Horses with the 3 Carts carried 13 Load of dung from Spear's and 1 from Cicilas. Between 1 and 2 I rode over to Bovington and there dined, and about 4 I rode from thence and carried my Son J[oseph] K[ingston] to Heffelton and my Father and Brother M[ichael] S[aunders] Went and We staid there till about 8 and then broke up All together and I got home about 9. Nanny came home just After dinner so was at Home before we got [to] Heffelton. Mr Aldridge was here in the Morning and paid me for some Vetches and Went round and saw the Work which J[ohn] Allen have done here. 'Twas a pretty deal of Rain in the Night but dry all the day and began rain again about 6 this Evening.

8 November.   I believe it rained all Night and I scarce remember it ever rained faster than this Morning about 6 for about of a Quarter of an hour. Brother M[ichael] S[aunders] appointed to Call me about 6 but he did not come till about 8 and he paid me £15 of £30 which He Owes me and 8s. for Interest, from July 9 to this Time. We thought to have rode to Blandford Fair together and I thought to buy some more Ews there, but the rain came on so that We did not go and Brother Staid here with us but the Wether was somewhat favourable in the Afternoon so After dinner I, Nanny and Bother M[ichael] S[aunders] Went up to

Affpuddle to Uncle T[homas] Saunders's and Edward Saunderses and came back again about 8. All the 9 Horses Carried with 3 Carts 11 Load of Dung from Spear's and fetched up 3 Load of Clotts to the Great Wear to make a Bay there.

9 November. Between 8 and 9 Brother M[ichael] S[aunders] departed. 3 Horses rested and 6 with the 3 Carts carried 7 Load of Mixoned Mud from the Mill pond and brought back 6 Load of Gravel into Fusgoalds to the new 2 Hatch Weir which Simon Hallet, T[om] Hunt, Dairyman C[harles] Hodges, I and John Antle began Setting off, We 2 went after dinner. We got in the Main Sill, the Staples and Head piece on but the place is very springy and the Gravel very hard to get up, 'tis much labour to lave and keep the Water down but We've got it in pure deep. I made 'em Sink it 6 or 7 Inchs deeper than they had put it in at 1st. Several showers in the Night and day but the Wether is very Mild. Nanny rode to Beer since dinner and bought some things to make S[tephen] W[hite] and J[ames] L[yne] each a Waistcoat. Several Showers again this day.

10 November. All the 9 Horses with the 3 Carts carried 9 Load of Mud More from Mill Pond *[space]* into the 6 Acres and *[space]* into Buckshill and 2 Load of Dung from by the Carthouse here at home and 1 out of Cicilas Barton and 2 from where the Mixon Was and *[space]* Load which they shot down where they broke the Draughts, in all 15 Loads. There is about 129 Loads in the 6 Acres which by Measure is but 4 [acres] 2 [roods] and 9 Load in Buckshill. I heard it rain Smart in the latter Part of the Night but 'twas a blessed fine, pleasent Mild dry day. S[imon] H[allet] and the Waterman got in all the New Wear, but the Landsides ben't quite all filled in. Galton and his Wife came about 1 and departed about 7. The Dairyman came in the Evening and Paid me his Quarter's rent which is not due till next Sunday.

11 November. 6 Horses ploughed with 2 Sulls and I held 1 in Buckshill 6 Acres and 3 carried over the Drag, a Sull and Harrows and Afterward Fetched in 100 fusfagots and then harrowed in the Wheat We sowed which was about 5 Bushels and it sowed all the East Furlong and 4 ridges of the Other. We cleaned out the Carthouse after dinner. A Mild pleasent day, the Lord grant several of them. I haven't sowed any Wheat before Since the Drot in Cockroad this day fortnight which is now coming up.

12 November [Sunday]. A smart shower about Sunrise, afterward fine and fair. My Father and son J[oseph] K[ingston] came and dined and went to Church with us. Text Proverbs 17. 3, and they Sung Psalm 4th. Henry Stickland of Bloxworth Offered to drive my Team. My Man George [Sinnick] Seems to be down in the Mouth. My Father and son departed about Sunset. I Lost another the 3rd Calf (Sue's) in the Kill Calf.

13 November. All the 9 Horses ploughed with 3 Sulls and sowed better than 6 Bushels of Wheat, which Sowed 11 ridges. I and Simon [Hallet] made a place in cart house for the Sow to Pig in. After dinner I and Jack Philps drove the other 2 Calves to Woodstreet and I came back by Bindon, the Roof of the Cowpens there is all up and the Thatchers are covering them. 'Twas a hoar in the Morning and a fine day. I got back about 8.

14 November. 8 Horses ploughed with 2 Sulls till about Noon. Then I had out the Other and 3 kept on ploughing, 3 Dragged and 3 harrowed and We sowed 6 ridges which took very little Odds of 4 Bushels. Mr Frampton was here about 11 and looked on the Wear at Fusgoalds and the Sluice in Gully which Simon Hallet, T. Hunt, Dairyman and J. Hodges began Setting in this day and finished it Except Gravelling the Landsides. J[ohn] Woodrow was here a Thatching the East Wall of the Garden and Mr F[rampton] had a long Talk with him about looking to Poachers as We Suppose for 'tis said that J[ohn] W[oodrow] has informed against 4 for Killing hares. Frost last Night. Close, Cloudy all day and Mild.

15 November. 8 Horses with 2 Sulls finished plowing the Wheat ground and ploughed a little in Buckshill they finished Sowing and dressing that field and brought home the Tackling to go on tomorrow again in Cockroad if please God. The Wether proves good, it rained a little in the Morning but not much, it broke away fine in the Afternoon. The Watermen finished Graveling the Sluice in Gully and began digging the Main (to let the Water into Horner's Mead) along by Gully Rails. I rode in the Forenoon and carried my Mother some Turnips and fetched between 2 and 3 Bushels of Wheat to sow, which Father had left of what he Bought. Mr F[rampton] and the Gardener rode by here about 8 in the Morning homeward. He said but little and I can't think where he had been. About Noon I went with John Woodrow to Hundreds Barrow Court and We dined at Beer. Nanny rode thither and bought some things. Robert Bower and John Hobs of Affpuddle and Hungerhill are Constables and Henry Willshear Tythingman of this place.

16 November. [shorthand characters] 6 Horses ploughed with 2 Sulls in Cockroad and sowed about 2½ Bushels Wheat there that I brought from Bovington yesterday, it sowed 4 Ridges. Simon Hallett and 3 Watermen set the Sluice by the South Bridge in Gully to let Water to Horner's Mead and he put up the South Bridge. Mr Frampton's Gardener and 4 Men began Banking in the East Barrow on Black Hill to set Firr Trees thereon. He talks of Planting more in the South Heath. A hoar in the Morning but fair and warm after.

17 November. 6 Horses ploughed with 2 Sulls and 3 Fetched in a Load of Fursefagots and Afterward 2 Went to harrow. I sowed about 5 Bushels and 3 Ridges more at the West Side and Missing 5 in the Middle it sowed all the rest at the East side Except the Headland, but the Ground is so Wet 'tis but dirty Work. The Watermen finished the New Main in Gully and Cleaned out some of the Old one and Graveled the Landsides of the Wear and the ends of the Bridge. I and Nanny rode over to Bovington After dinner and Staid till about 8. A little rain in Morning, dry and Cloudy Afterward but we had a smart shower coming home.

18 November. 6 Horses at Plough and 3 at Harrow and Afterward 4 at Plough and 5 at Harrow. Made an end and finished plowing and sowing of Cockroad, carried out 7 Bushels of Wheat and sowed very near all of it. There is about 39 Bushels sowed in that field and about 19 Bushels in Buckshill 6 Acres. I sowed till about Noon then came home and rode to Wareham Market. Corn seemed at a stand. There was a Great many Loads of Corn at Town and several Farmers Eastward and from the West. Showry wether most of the day. I saw Father White at Market and was in at Brother F[lorence]'s. I sent a letter to Mr Brander at Nea and another to Dr Bright at Ringwood. I got home about 7.

19 November [Sunday].   It rained I believe most of the Time from sun rising till Sunset but Mr Fisher came and about 17 or 18 more and I and Thomas Willshear Sung with the Clark, W. Spear, Joseph Boyt and Joseph Snook. The 1st Time I ever Sung with Puddle Quire. Text Ecclesiastes 12. 7 and we sung Psalm 5th. That [?fond] Wench Ann Hodges was up at Stable talking with my Men till [al]most dark and so she have, they say, Several Sunday Evenings before.

20 November.   All the 9 Horses with 3 Carts Carried dung into Hilliar's Plott. Mr F[rampton] called here and seemed Angry because I had not got the Old Brix together and said he would send his Own Teem to do it, but I promised him to do it tomorrow. The[n] He asked me what I Thought to do with my Fir Trees at Bovington and I sold the whole Bed of my last setting to him for 5 Guineas. He was then going up to see the men at Work at the Barrow on Blackhill and As he came back, He left word I should make an End Sowing first before I did carry the Brix. My Man John Antle Threshed about ½ Bushel Oats. He walked away out of Barn and did not come home to dinner. I mus[t] Question him about this. Mr Vanderplank, Mr Bartlet and John Whennel was here and spent the Evening with us till 10. Som few drops but dry for the most part.

21 November.   I asked my Man John [Antle] whither he went to yesterday but He would give me no Account so I set him to fill dung today instead of Threishing. All the 9 Horses carried dung again into Hilliar's and [?poor] George [Sinnick] had a Misfortune to hurt the young mare Whitefoot, in the sid of the Face under the Eye there is a round hole broke into the Bone 2 or 3 Inches deep. I think 'twas with a punch of the Whip But if so he denies it and said she fell against the Post coming out of ground but I can't perceive how such a hole should come if she did fall. He pretended he did not know of the hole till some of 'em saw her bleed at Nose. I rode to Bovington in the Evening and borrowed 19 Sacks. I with R[obert] P[hilps] and S[imon] P[hilps] Winnowed 8 Quarters of Barley which R[obert] P[hilps] Threished by the Quarter at 10d. and Earned about 7d. a day and so he did a spurring of Dung at which I gave him 1s. for the Lug Acre. Wind at North and dry.

22 November. All 9 Horses and 3 carts carried dung, Mud and dirt and what they could rise to finish Hilliar's plott. I, S. P[hilps] and J. H[odges] Winnowed 13 Quarters and 6 Bushels of Barley 10 of which I intend to carry to Bere to Morrow. A Calm, Foggy day some few drops of rain in the Evening but soon broke away. The dairyman sold his Colt for £4 14s. 6d. at Martins Town fair and got home by time I and C. H[odges] went and Looked out some Wood to make Cow Cribs. I dressed the Mare this Evening and gave George [Sinnick] Notice to get him a place by St Thomas's day *[21 December]*.

23 November. George [Sinnick] and J[ohn] A[ntle] carried old Brix together with a Cart and 2 Horses and I and Tom [Hunt] with the Wagon and 6 Horses carried 10 Quarters of Barly to Beer to Jay Burges and as the Wagon came back brought a Load Frith out of the Copice by the Cowlease to make Cow Cribs in Dairy barton. And C. and J. H[odges] the Dairyman and T. Hunt made up the Cribs and *[space]*. After We unladed at Beer, I rode to Bloxworth and spoke with Henry Stickland and he'll be here tomorrow to see a house and bargain if we can. I called at Mr Ekins's to see if he had any Bushes to sell and after dinner I rode to Bovington and carried home my Father's sacks. Fine, dry Mild Wether.

24 November. We ploughed with 9 horses and 3 Sulls and Sowed about 6 Bushels Wheat at Hilliar's and it Sowed *[space]* Ridges and *[space]*. H[enry] Stickland came but We did not bargain for he would not come under £15 12s. which is 6s. a Week all the year round. I Offered him £15 and 1s a Load for every Load of Lime He should fetch. John Harding cut our little Pigs. He told George

[Sinnick] of a place at Farmer Hayter's at Langton by Blandford. Fine drying Wether.

25 November.   All the 9 Horses finished plowing and Sowing Hilliar's Plot which took about 13 Bushels and 3 Peck. Close dry Wether. George [Sinnick] went to Blandford to see about a place, but did not hire himself. I rode to Wareham Market. Corn Seems to fall, Barley at 21s. and 22s. [a] Quarter. My Father was there his Wagon carried a Load. I did not Sell any. I was in at Brother F[lorence]'s and Sister told me that she was at Heffelton yesterday that Father White came home while she was there from Martins Town fair and was not well, was Ill of a Cold and that the Newsman Told Mother White that Brother Brown's goods at Martin was seized and sold for selling Spirituous Liquours without Licence and that they was again at Handly.

26 November [Sunday].   Mr F[rampton] sent R. White to Invite me and Nanny to dinner and we went. Prayer was just in before we got to Church. Text Colossians 4.12 last part and they sung Psalms 23, 18ª. The Buttler gave us some Mulled Wine before dinner. There was only Mr F[rampton], Mrs [Frampton], my Wife and me at dinner together. He have drawn up a scheme for my lease and was willing I should hear it read before he did send it to Mr Humphrys to have it engrossed. He gave consent that I should Get more Fir seed off his Trees at Moreton. We came away about Sunset. 3 Houses, a Barn, a Stable and a Carthouse was this day Burned down at Anderson. The Wind was High. Text here Ecclesiastes 12. 7 and the Clark sung Psalm 142.

27 November.   3 Horses carried abroad Earth in the flat by Robert Philp's and 6 ploughed with 2 Sulls, finished Buckshill and done about ½ the Flatt and sowed almost all that is ploughed. Nanny rode to Heffelton to see her Father who is ill. I held sull all day and was going to Heffelton after dinner but met Nanny at Gallows hill. My man George [Sinnick] went home yesterday and did not come till night. Tom Hodges went to Bloxworth and from thence to Anderson on the Alarm of the Fire and 'twas [al]most Sunset when he got back. They both went away unknown to me, But John Antle Asked me and I denyed him but he went to his friends at

Possham and is not returned yet past 8 at night. 'Twas a Close, Cloudy, Cold day.

28 November. 3 Horses rested and 6 made an end plowing and sowing of the Vetches in Bucks hill Flatt. I rode over to Beer and received of Mr Burgess the Money for the Barley I carried him last Thursday and Afterward rode to Bovington and carried my Mother some Turnips and to Bindon Where was my Father with his Carts carrying Gravel into the backside before the New Cow pens. I just called and saw Tom Barratt and John Elliott and Offered John E[lliott] some Barly but he would not give me but a Guinea a Quarter. Mr B[urgess] paid me 22s. a Quarter and so I offered it to John. I rode from thence to Heffelton and saw Father White who is Ill and looks Very Ill. From H[effelton] I rode to Bovington and staid a little with my Father and Mother and Father gave me an Account of some Butter sold to Ringwood, the Mony of which I am to receive tomorrow. My Man John Antle came about 2. He is now out in Stable but I haven't seen him. About 8 soon After John came in I paid him 6s. 1½d. and Bade him begone and So he Packed up his Things and departed but would have Staid. He went and lodged at John Rawls's.

29 November. 4 Horses and 2 Carts carried Clots and Gravel and made up the bank of Rack mead main and 5 Horses and 1 Cart carried together more Old Brix and 2 Load of Batts to the Fords in Mead and 1 Load to Hilliar's Barrs. I Set out about 4 in the Morning for Ringwood, got to Pottern *[in Verwood]* about 8 and between 9 and 10 rode from thence with Uncle and A. Duel to Ringwood. Took some Butter Money there for my Father and Expected some of Mr Brander but he was not there nor had not left it. I dined at Dr Bright's and went and saw all Cousin James Warn's Family and Cousin Thomas Russel, Henry and Sarah Jessop and about 3 set out for Nea and Lodged at Mr Brander's.

30 November. 3 Horses rested and 6 fetched a Load of Gravel to the Way by Rack Mead and 1 Load of Frith from Buckshill to the Home Barton and a Load of Furse out of Brick Close. Mr Brander told me He had wrote to Mrs Shaw to Pay My Money to Brother Loxly. So I came from thence between 4 and 5 and through

106

Christchurch and by Poole and to Litchet by Sunrise and got home between 9 and 10 and Taking a little Refreshment and the Old Mare I rode to St Andrew's Milborn Fair. I saw my Father there, He did not seem to be very well and came away before me. I bought 20 Young Ews there for £10 12s. od. of Farmer James Loveless of Colliers Puddle. A Frost in Morning. Fine Wether but Cold.

1 December.   6 Horses and 2 Carts carried away Mud out of the River between Brockhole Mead and the New Bay and 3 Horses with 1 Cart fetched more Gravel to the Way and Bank by Rack mead. I filled some Mud and mended the Dairy Barn's door and went round the fields and saw the Cattle. Since Dinner I've been at the Blacksmith's to bespeak some things and came back round by Shitterton and thought to have Took Alner's house for Henry Stickland, with whome I bargained yesterday at Milborn and am to Give him £15 12s od. for a year and I to have all the Lime that We fetch with out Allowing him anything for it, and we Appointed Monday senit next to fetch him. But Although Old Talbot proffered me the house, Garden and Orchard but last Tuesday for 20s. a year yet now he have considered of it and won't let it at all. A Frost fair and Cold.

2 December.   3 Horses carried Gravel to the Great wear and 6 with 2 Carts carried more Mud out of the River. I rode to Wareham Market and carried 8 lb. Butter and Sold 2 Load of Barly at 21s. a Quarter to Mr Phippard, to carry 1 Load next Tuesday and the other about Friday. I did not hear of any sold for any more. I heard my Father was gone to Dorchester. My Dairyman has been here this Evening and asked me if He should go on for another year. I am Willing He should and proposed milking 30 which he is as Willing to Take as 20, If we can Suit ground for 'em. A Frost and very cold raw Fog.

3 December [Sunday].   I and Nanny Walked to Beer Meeting in Morning. Text *[space]* and they Sung Psalm 33 and We went to Church in Afternoon. Text 1 Timothy 1, had no Singing. The Wether very Cold and but very few people. Yesterday was 7 year ago I and my dear Spouse was Married Since which God has been pleased to bless us with 4 Sons.

4 December. Snow and rain when we Got up. 3 Horses rested and 6 fetched a Load of Timber from Bovington for the Wear by Cicilas Mead and Plank for Fusgoalds and the Great Wear and a pig's Wick. I rode to Bovington and to Woodstreet and Bindon and helped up some Cattle with my Father, to Take to fodder as he did all at Bovington this Evening and tomorrow do intend to fodder them at Bindon. Their New Pens are finished and tomorrow will finish graveling the Barton. I called at Heffelton as I came back and saw Father and Mother White who are both better.

5 December. It rained when we got up and so Continued till Night. 6 Horses carried a Load of Barly 8 Quarters to Wareham and brought back 60 Deal boards to make Hatches and Barns door. A Cart and 2 Horses carried a few Old Brix together, but it rained so fast they shut out. I was in at Dairy house and the Dairyman Told me he was sent for yesterday about a dairy of 44 Cows and as We Talked abut our Cows and the Ground He Said I should let the 20 have most of the Ground I did Talk of for 28 or 30. I believe I shan't let above 27 or 28. We did not resolve or conclude on anything for I think We need not be in a Hurry 'tis a long time to Holliry day.

6 December. 5 Horses fetched a Load of Frith from Buckshill to Spear's Barton to make Cow Cribs and fetched the Fold from thence up into Saltfield and a Cart and 2 Horses carried the last of the Old Brix together and We covered 'em. Then All the Horses and 3 Carts carried 6 Load of Batts down into Mead to make good Fords there. A little Frost in the Morning and the Wind at North blew cold all day but fine and Clear and very drying. I rode to Beer Market and bought some Meat and Candles and wrote a Letter to Brother Loxly at Martin and left it there to be carried by the Newsman. I rode from thence to Bloxworth and Told Henry Stickland, Old Talbott would not let me that House. The Gamekeeper and R. White was round here a shooting and shot a Fox, 3 Woodcocks and 3 Rabets. The Rabets they brought in and gave to Nanny, which was very kind.

7 December. 4 Horses rested and 5 carried most all the rest of the Brick Batts to the Fords in Mead. I Winnowed Barley. Mr F[ramp-

108

ton] came to Settle about what He Should allow for Church and Poor Rates. He Seems Willing to Allow but £8 and it comes to £8 16s. 9¾d. a year. He Seems to be in a pretty good Humour and we Told him about a Court before the Door and he said it should be pitched and He consented I should let Henry Stickland into the House at Spear's. A sharp Frost, I think it Froze most all the day in the Shade. Carpenters and Watermen began Setting the Sluice going into Cicilas Mead.

8 December. 3 Horses rested and 6 drew the Fan Tackle to Dairy Barn and with 2 Carts fetched 4 Load of Gravel to Cicilas Wear a load of Clots and carried away 4 Load of Sand and dirt and brought up the Chaff from Dairy Barn, a Brave day's work. I'm provoked to see how Slow they are. I, Simon P[hilps] and J. Hodges Winnowed Barley at Dairy Barn and they have Laded it this Evening to carry it to Wareham tomorrow. S[imon] Hallet and Andrew *[space]* finish Cicilas Wear and put up the Stile by it. A Smart Frost but Thaw a little in the sun. Nanny's very Ill of a Cold. Cousin W. Fry fetched a Pott of Butter.

9 December. George [Sinnick] with 3 Horses carried away Rubble from behind the house into Brick Close and 6 carried 8 Quarters Barly to Wareham which was 1 of the Load I sold last Week and I received the Mony a Guinea a Quarter for it. I would have sold another Load but the Malters don't care to Buy 'tis quite dull work with them. My Father was there and his Teem of Woodstreet with a Load of wheat. No frost But a mild pleasent Thaw, small rain this Evening.

10 December [Sunday]. Rain most of the day. I went to Church in Afternoon. Text 1 Timothy 1 and the Clark sung by himself Psalm 3. Mrs Willshear and Mrs Woodrow came in and drank Tea in the Evening.

11 December. About 4 I set out for Ringwood fair and got there between 8 and 9. I bid George [Sinnick] carry Rubble with 1 Cart and 3 Horses from behind the House into Brick close and Tom [Hodges] to go with 2 Carts and 6 Horses and carry more mud out of the River but he went not, but carried rubble with George.

They had 2 Carts and 4 Horses and t'other 5 bade in Stable all day, such dutiful servants have I! I bought 21 Ews at Ringwood of Mr Cockram he to deliver them home for 10s. a piece, but they cost him in Purbeck 10s. 6d., they sold druggish there and Manny was drove home unsold. I saw several of my Kindred there and Brother Loxly and went and Laid at Farmer Brixey's at Upper Kingston. As I was riding through Beer Field, about 4 in the Morning I saw a Great Star Shoot in the East which Appeared very bright and a stroke where it shot continued a good while after. 'Twas a fine day.

12 December. Farmer J. Brixey and I went and breakfasted at his son's with he and his Wife and Farmer Richard Bushell of Munkton and about 11 I came from thence through Ringwood and called at Cousin J. Warn's and Mr Moody's the Cutler and staid and dined at the Poor house with Cousin T. Bungy and his Wife and there Took a Copy of the Rules of the House. I came from thence about 3 and about Sunset called on Cousin Joseph Effemay at Lye at Wimborn and staid with them about of an hour and got home between 8 and 9. 3 Horses rested and 6 and 2 Carts carried mud out of River, the fellows was loath to go but the dairyman came up and they went down with them. Several Showers and brisk Wind.

13 December. I set George [Sinnick] and Charles [Hodges] with a Wagon and 5 Horses to fetch home more Frith from dairy Coppice to make Cow Cribs and bad[e] Tom [Hodges] go with the Carts to Carry Mud to River but he would not go and so said George. So I and dum Jack had down the 4 Horses and the 2 Carts and Jack and Old T. Hunt filled and I a little and the Dairyman drove it. And After we was gone Tom Took his Cloths and set out, about Noon after I had got Farmer Willshear to Turn the Water to make it low, I rode to Beer and bought some things and Called on Shepherd Coaks and talked with him about his Son, to be a Shepherd, which he said should come over and talk with me within this week. From thence I rode to Bloxford *[sic]* and Told H[enry] Stickland I had got the loft new laid for them and appointed to send the Waggon to fetch him next Friday, or if it should rain then, to fetch him Monday and Then I rode to

Heffelton and saw Father and Mother White who are both on the mending hand and from thence I rode to Wool, thinking to have bargained with John Notting but Met with my Father there who told me he dwelt [?dealt] with him, so I went with my Father to T. Barratt's and came back with him to Touthill Gate. There Farmer P. Phippot of Beer overtook us and I came along with him.

14 December.   S. Philps and his Son ploughed in Longcraft with 4 [horses] and George [Sinnick] drove out mud notwithstanding He said yesterday he would not. He was 'fraid If he did not work I should not pay him and did not refuse going we had 2 Carts and 2 Horses to Each and 1 Horse rested. Dairyman, T[om] Hunt and J. Hodges filled and I was with them and filled some. The Carpenters finished the loft at Spear's in the Evening T. Hodges came up to see if I would pay him or set him to work. I told him as he went yesterday for his pleasure he should go to Night for mine and come up tomorrow Morning with his Father and then I would talk with them together. A frost in the Morning, dry, fine day and not very Cold.

15 December.   Tom [Hodges] came this morning with his Father and I reprimanded him pretty much. I told him if He would go to Mud cart and do what I had for him to do and behave as he ought He might go to work, but he said then he would not go to Mud Cart, but his Father Chid him and he considered of it and went with George and T. Hunt and the Dairyman to mud Cart with 2 Carts and 4 Horses. Sent S. and J. Philps with the Wagon and 5 Horses to Bloxworth to fetch Henry Stickland and his Family, which arrived between 1 and 2 and his Father's Wagon brought him a Load of Fewel. Farmer John Mate called here about 10 and staid till 12. Since dinner I carried my 4 Boots to Beer to have them mended. A Close, mild day. Shepherd Coks *[sic]* of Beer came this Evening and agreed with me for his Son Luke to Come Tuesday next for 45s. 6d. till Michaelmas next.

16 December.   H[enry] S[tickland] with 5 Horses fetched the rest of his Goods and 4 with 2 Carts carried Mud out of River. They carried all just above the Wears below the rails and some above out into Fusgolds. I rode to Wareham Market and bought Some

Iron Shovels there to fill it. Barly is falling more and more, I heard there was some sold for 17s. [a] Quarter. I paid Mr Cockram £10 10s. od. for the Sheep I had of him at Ringwood fair and I paid Mr Brixy for Stone Tile had here about the house. My Father's 2 Wagons carried 2 Load Barly but he was at Dorchester Market. Mild plesant Wether.

17 December [Sunday]. I took Physick last Night and this Morning which worked 4 Times. Between 10 and 11 Joseph Davison of Highwood and his Wife Susan came to see us. 'Twas a Mild Morning but rain all the Afternoon so he went home about Sunset and left her here. About Noon Cousin William Tilly, Junior, called as he was going to Home but would not alight. One Homer of Rogershill a young man called, he heard I wanted a Servant and before we went to Church Joseph Applin of Shitterton called to ask for work. I'm like to have work folk plenty. We went to Church in Afternoon the Clark Sung Psalm 37. No Sermon but Mr Fisher Expounded on the Exhortation to the Holy Communion.

18 December. H[enry] S[tickland] with 4 Horses ploughed in Long Craft and 5 with 2 Carts carried out more Mud in Fusgolds. I rode to Bovington and Carried my Mother Some Turnips and from thence to Woodstreet where my Father was winoing some Raygrass Seed and with him I came back by Bindon. There the Labourers was making a Hedge round the Barton and Mr Weld has sent some Ash and Beech Trees which are Planted round the West and South Sides of the Barton, which Shews Neat and the Barton appears to be dry and healthy, my Father having had much Gravel carried into it and it being well Bedded up, with the New Cow Stalls at the East sets it Off very much and it has quite a New Aspect. I got home between 4 and 5 and soon after came Brother S[tephen] W[hite] and brought me £12 Interest Money from Mr Brander, which Brother H[enry] L[oxley] received at Sarum and sent from Blandford last Saturday by Brother S[tephen] W[hite]. Cloudy but very Mild.

19 December. Brother S[tephen] W[hite] laid here last Night but departed Early this Morning about 7. H. S[tickland] with 4

Horses ploughed Long Craft and 4 with 2 Carts carried out the rest of the Mud in the River above the Rails in Fusgolds. 1 Horse rested the Forenoon and afterward drew out Chaff from Cicilas Barn into the 7 Acres and brought home the Dust and Fan and C[harles] H[odges] and S[imon] P[hilps] Winnowed 4 Quarters 6 Bushels Oats. 'Twas a blessed fine, Mild day. Nanny continues very bad. I went to Beer in the Morning to Dr Best's for her but the Doctor was not at Home. She seems a little better this Evening. Cousin Nanny Langford of Bere came to see her this Afternoon and is here now. Luke Coaks, a Lad of about 17, of Bere came to live here as a Shepherd for 45s. 6d. till old Michaelmas.

20 December. Miss Manuel was Married yesterday at Beer to Mr Sparks and her Mother died in the Evening. H. S[tickland] with 5 Horses fetched a Load of Timber from Bovington, 2 Single Sluices, Several Bar and Stile Posts and a Load of Fus for himself and the other 4 with 2 Carts carried dirt and Clots out of River at the lower End of Lilington's Mead to fill up the Old Main in Fusgolds and fetched some Gravel for the Sluices in Cicilas Mead, 1 of which John Allen and his men set in. Finished the Great Fishing Wear and put 2 Posts to the 7 Acres. I rode to Beer Erly in the Morning and had a Blister for Nanny who is very Bad and her Disease does very much Affect her Eyes, so that she can't bear to look on the light. After I came back between 7 and 8 I rode to Bovington and from thence to Wool and there Mrs Knapton of East Lulworth lent me £60 on my Own Note of hand and I am to give her 4 per Cent for it. I got home about 3. A Fine Mild day, a little rain.

21 December. 2 Horses rested. H. S[tickland] with 3 [horses] and 1 Cart carried out dung and dirt of Spear's Barton and brought in Chalk to make it plain and 4 and 2 Carts carried more Dirt and Clots to fill the Old Main in Fusgolds. I Paid Off George William Sinnick this Morning and he is gone. My wife continues Bad, she had on her blister last Night and it rise very Well. Cousin Nanny Langford is here with her yet. Brother R[oger] White came to see her about Noon. Some rain in the Night but very little all the day, very Mild and pleasent. J. Allen and his Men put in the Other sluice in Cicilas Mead, 3 Bar Posts, a Stile and something to the Foot Bridge at the Lower End of Fusgolds.

22 December. 2 Horses rested and 7 with all 3 Carts carried Chalk, Gravel and Clots to make a Rolling Bay at the New Bay going into Brockhole Mead and some Clots in Cicilas Mead. J[ohn] A[llen] and 2 Men finished the foot Bridge by Cicilas Mead, put a Stile between Ten Acres and Brockhole Cowlease and plained Boards for Book shelves. It began to rain about 11 and continued till [al]most Sunset. After Dinner We moved out Corn from the Little Chamber and Got out the best Bed from the Great Old Chamber and put it up in there. Had 2 Lams this day. The first was a Ram Lamb, which was Lambed in the Night. 'Tis now a fine pleasent Evening. Nanny Continues Ill.

23 December. All the 9 Horses with 3 carts carried Chalk, Gravel and Flints to the Rolling Bay and brought some Straw from Cicilas Barn to the Old Barn at Spear's and brought over the 10 Cows that was over in 10 acres into Spear's Barton and Fodderd 'em there. I rode to Wareham Market carried to Fat Geese, Corn low, Barly about 15s.[a] Quarter. I saw my Father and Father White there. A fine day. I got home between 7 and 8. J[ohn] A[llen] put up Book Shelves in Kitchin and made Hatches for the 2 Sluices in Cicilas Mead. Had but 1 Lamb last Night and that was Dead.

24 December [Sunday]. Small drisling rain in Forenoon. My Father and son J[oseph] K[ingston] came and went to Church with me. Text 2 Corinthians 5. 10 and the Clark sung Psalm 108. I, W.C., W.D., G. and M.W., Mrs W., W.S., and T.H. staid and received the Blessed Sacrament. After dinner about 1, Brother M[ichael] S[aunders] and his son M[ichael] came. I left 'em here and rode to the Funeral of Mrs Manuel. They Gave me a Black Silk Hatband and White Gloves for me and my Wife, we set out from Chaleys between 2 and 3 and carried and Buried her at Bloxford *[sic]* she was 58 years Old. There was Many of their relations and Others and most of us that was invited to Chuly house went After we come out of Church and Eat and drank at the house of young Mr Manuel at Bloxworth, we got home about 7. Brother M[ichael] went to Bovington with my Father.

25 December. I rode to Bovington Early and fetched a Lamb to set to the Ewe that lost hers on Friday and Luke C[oaks] set it. I rode to Beer Church in Forenoon. Tex[t] Isaiah 9.3 last part and they Sung an Hymn and Anthem out of Luke 2nd. I, the Exciseman Mr Moors and his Wife and his Brother the Schoolmaster and Farmer John Woolfrys Dined with Mr Fisher and Staid till about 7 in the Evening. A Cold Hazey Air Some Times the sun Shone Out. There was a pretty many People at Beer Church. Yesterday while I was at Mrs Manuel's Funeral, Brother S[tephen] W[hite] brought Sister Patty and left her here and carried away my son J[ames] Line with him.

26 December. The Dairy Man had a Cart and 2 Horses and his Own Mare and carried our 5 Tubs of Butter and 6 C[hurns] of his Chees to Wareham. H. S[tickland] ploughed with 3 [horses] in Long Craft and T[om] H[unt] carried Gravel to the Sluices in Cicilas Mead and the Dam. About 9 or 10 Folk from Beer came and paid for their Turvs. They use to have a Feast for them but [as] Nanny Continues Ill, we had only a Cold Ham and Bread and Cheese and Butter and what Ale and Cyder they would drink which pleased most of 'em very Well. They Bragged Mightily on the Cyder and drank very hearty of it.

27 December. H. S[tickland] Finished plowing Long Craft with 4 Horses and the other 5 with 2 Carts dug out a way in Cicilas Mead to make a drain of it and carried some Chalk and Gravel to the Rolling Bay in the Dam and the Suls they brought Home. I rode to Bovington about Noon and carried my son S[tephen] W[hite] with me. Brother M[ichael] S[aunders], his son M[ichael] and T. Pearce was Fereting Rabits in the heath and I went to 'em, they Killed 9. I Staid there till about 9 Clock and left S[tephen] W[hite] there. Brother Michael and his son Intend to go home tomorrow. A sharpish Frost in the Morning but fair afterward.

28 December. 7 Horses and 3 Carts Carried gravel into Fusgold's for the Great Wear and some to the Rolling Bay. About Noon Father White came to see us. S. Young Carried a Fat Goose to Heffelton and Bettsey Florence came from thence with her and in the Evening when Father White rode home Sister Patty rode home with him, on the same Horse which Betsey rode hither. Close cloudy Wether with some small rain. I had 3 Men, Thomas and George Hunt and the Dairyman digging a New Main Through Cicilas Mead to Water Spear's Moor that way and I hope 'Twill answer very Well. Nanny continues Ill, very little better.

29 December. 2 Horses rested and 7 with 3 Carts carried more Gravel To the Rolling Bay and some they Shot down by to rise the Bank from the Rolling Bay to the Great Wear. C. H[odges] cut down and rid away Alders from the brink of the Ditch in Spear's Moor by Mr Manuel's Mead and the Dairyman and George Hunt routed out all the Mores by that ditch after him and T. Hunt Took up the Edge of the Ditch after them to Widen it to make a good drain of it. A little rain in the Night but a blessed fine day. This Evening have 13 Lambs.

30 December. H. S[tickland] with 5 Horses fetched a Load of Frith, a load of Mall and a Load of Mores out of Spear's Moor and 4 Horses with 2 Carts almost finished filling the Old Main in Fusgolds. I rode to Wareham Market and Weighed my 5 Tubs of Butter which I sent thither last Tuesday. The Dairyman Drove 2 of our Barton Pigs to Wareham Market but did not sell 'em, but 1 of 'em Tired and the Other he lost and the Maid left Open the Turf house Gate in which stands the Wash Tubs and 2 of the little pigs got in upon the Turf scroff and fell into the Tubs and was drowned, they was worth 4s. or 5s. a piece. Mild weather.

31 December [Sunday]. 'Twas a Hard frost in the Morning but about Noon turned to Rain and rained till Night. Cousin Hannah Fudge and her Daughter Nanny Langford came to see my Wife who God be praised is something better, Nanny and Betty Florence went to Church and set with me. Text Matthew 5. 20 and they sung Psalm 106. After Prayer Mr Fisher gave away several Loaves of Bread as Usual at this Season to the Poor People. He would have come in and staid here had not it rained.

LETTERS FROM GEORGE BOSWELL OF PUDDLETOWN TO GEORGE CULLEY OF FENTON NEAR WOOLER, NORTHUMBERLAND, 1787-1805

INTRODUCTION

GEORGE Boswell was born in Norfolk in 1735, but came to Dorset as a young man and spent the rest of his life there, with only occasional visits to his native county. His letters reveal little about his background, or about the reasons why he settled in Dorset, but it seems likely that he came to Puddletown in the service of Robert Walpole, second Earl of Orford, whose principal estate was at Houghton in Norfolk. The earl had acquired the manor of Puddletown through his marriage in 1724 to Margaret, daughter and heiress of Samuel Rolle of Heanton, Devon.[1] There were several other families named Boswell living in Puddletown during the eighteenth century, but there is no indication that George Boswell was connected with them.[2] According to his own account in his letter of 14 March 1787, Boswell quarrelled with an aristocratic patron, possibly the earl of Ilchester, who possessed lands at Woodsford in the Frome valley south of Puddletown, and to whom Boswell had dedicated his book on water meadows in 1779. Thereafter Boswell seems to have engaged in farming and other business enterprises on his own account.

On 12 June 1760 he married Sarah Chapman of Puddletown by licence in Puddletown parish church, and for much of the rest of his long life he lived in a house in the Old Market Place at Puddletown.[3] His letters are all headed 'Piddletown', although during the 1790s he was living at Waddock Farm in Affpuddle. In the survey of the inhabitants of Puddletown made by the vicar, the Reverend Dr Philip Lloyd, in 1769 George Boswell and his wife Sarah are shown as having a large establishment, for their household in the Old Market Place included three children, four servants and two of Sarah's unmarried sisters.[4] George Boswell's wife, Sarah, was the daughter of the Reverend John Chapman of Puddletown, who was a schoolmaster and also took gentlemen's sons into his household.[5] Several children were born to George and Sarah Boswell, including George 1761, Sarah 1763, Robert 1764, Sarah 1765 and Samuel 1771, but several died at birth or in infancy, leaving only Samuel to succeed his father. Their mother, Sarah, died in 1777 at the age of 45, and thereafter George Boswell remained a widower until his death at the advanced age of 80 in 1815.[6] From his letters it is evident that after his wife's death his household was managed and his children cared for by his sister-in-law, and after her death in 1796 by his daughter Sarah.

Another connection between Puddletown and Boswell's native county of Norfolk was through the vicar of Puddletown, the Reverend Dr Philip Lloyd, who was dean of Norwich and also a canon of Westminster. He was vicar of Puddletown from 1765 until his death in 1790.[7] During the 1770s and possibly earlier, George Boswell acted as agent for Dr Lloyd in the collection of tithes at Puddletown, in dealing with the numerous and complex tithe disputes, in attending to the vicarage and the glebe lands, attending to the vicar's affairs in Puddletown, sending goods to him in London or Norwich, and keeping him informed about local affairs, since the pluralist vicar spent much of his time in London or Norwich. Later, Boswell leased the tithes of Puddletown from Dr Lloyd, paying him an annual sum.[8] Philip Lloyd obviously thought highly of Boswell and in August 1774 recommended him to Earl Temple of Stowe, Buckinghamshire as a suitable man to value the estate at Eastbury in the parish of Tarrant Gunville which the earl had inherited from George Bubb Dodington. In his letter which is now part of the Stowe Collection in the H.E. Huntington Library at San Marino, California, U.S.A., Lloyd goes on to describe Boswell as follows:

> He is the Mercer of our Town; & has in his own Occupation about £200 a year, partly his own, & partly hired; so that he is well acquainted with the Nature of

Dorsetshire Land, & Husbandry, & how far it may be improved. He has too some years ago acted as Manager of Lady Orford's Woods for his Uncle, who was the Bailiff of them: so that he is well qualified to estimate the Estates.[9]

George Boswell was also involved in other enterprises, such as representing James Frampton of Moreton and other local landowners in various dealings over enclosures on the heath, forestry and land reclamation. He describes himself in Letter 1 as 'a Tradesman and a Farmer', in Letter 13 as having a 'grocer and mercer's business', and in Letter 17 as a maltster. In the 1769 survey one of the buildings in Puddletown is described as 'Mr Boswell's Skin House', so he may have been involved in tanning in addition to all his other business affairs. Evidence of his involvement in trade as a mercer is provided by the Puddletown overseers' accounts for 1779-80 when they paid George Boswell the sum of £1 19s. od. for supplying bedding, flannel sheeting, tape, buttons and thread. Further sums were paid to him during the 1780s.[10]

His main interest was in agriculture and in the dissemination of improved methods and ideas, especially in spreading knowledge of the techniques and benefits of watering meadows. Throughout his life he remained a practical farmer, renting lands in Puddletown and district; he held leasehold and copyhold lands on the Earl of Orford's estate, and several parcels of recently-enclosed land on the heath at Affpuddle and Pallington, which were part of the Frampton estate.[11] He was obviously well respected as a farmer, and James Frampton was pleased to grant him a lease of Waddock Farm in Affpuddle upon very reasonable terms. This was one of the best farms on the estate, with fine new barns, a rebuilt and enlarged farm house, and extensive water meadows along the river Frome. In his Memorandum Book for *c.*1790 James Frampton recorded that

> . . . [Waddock Farm] has been in most slovenly hands and under sad management for a number of years, but is now Lett to Mr George Boswell of Piddletown, who entered on it and has conducted it with uncommon spirit, when much may, and I dare say will be done, if he

lives, whence some considerable advantage at the end of his Term may accrue to the owner. But here I will say, once and for all, that my opinion is, if rents are advanced, which I think but reasonable they should, agreeable to the Times, I would ever recommend it to be done with Discretion, keeping rather below than above the market price, and with great allowance always and indulgence to active spirited servants, particularly to such an uncommon one as Mr Boswell, whence the Landlord has many advantages, especially a choice of sensible, active, responsible and good tempered neighbours.[12]

There could hardly be a warmer tribute from a landlord to a tenant. In Letter 4, written in 1788, Boswell also refers to his land at Park, possibly Park Farm at Burleston on the river Piddle. In 1805 George Boswell and his son, Samuel, also took a lease of 125 acres around Oakers Wood on the Frampton estate for twenty-five years, undertaking to build a farm house, barn and stable. The barn and stable were duly finished by 1807 and the farm house in 1809.[13] This was the land which Boswell refers to in Letter 17. Oakers Wood had been cut down during the early 1790s, and the landlord, James Frampton, hoped that George Boswell and his son, Samuel, would be able to turn it into productive agricultural land, but in spite of all their efforts, the experiment was not a success, and after George Boswell's death in 1815 Samuel surrendered the lease and the land was once more planted as woodland.[14]

It is clear from his letters that George Boswell had received a good education, and that he had a lively, enquiring and essentially practical mind. He was obviously fascinated by the advances taking place in English agriculture, and by the new agricultural machines, such as threshers and drills, but he was also confident of his own ability to make them himself. He was involved in the management of the charity school at Puddletown, served as Overseer of the Poor in 1775, and was deeply concerned about the appalling conditions of the Dorset agricultural labourers. It is clear from these letters that low wages and poor housing were already causing discontent among the labourers during the last

120

decades of the eighteenth century, and that there was already a fear of riot and unrest, although conditions were to get much worse during the nineteenth century. During the Napoleonic War, when there was a threat of invasion, Boswell was named as one of the volunteers for driving livestock and taking charge of the evacuation of the district.[15]

His national reputation as a leading exponent of agricultural improvement was established in 1779 with the publication of his book *A Treatise on Watering Meadows*. This was dedicated to the earl of Ilchester and to James Frampton, and provided a full account of the methods and benefits of watering meadows, with detailed descriptions and diagrams of channels, drains, hatches, weirs and trenching tools. Boswell stressed the practical nature of his book and poured scorn on the work of other writers on the subject:

> . . . not the effusion of a garreteer's brain, nor a Bookseller's job, but the result of several years experience; and now committed to the press with not a little labour, at the request of some friends, as remarkable for their assiduous attention to the good of the community, as for their knowledge of the subject.

The book achieved considerable success, and a second, larger edition was produced in 1790. It brought Boswell to the attention of the leading agricultural writers of the day. It also brought him into contact with George Culley of Northumberland to whom this series of letters describing Dorset agriculture is addressed.

George Culley was born in 1734 at Denton near Darlington in County Durham, where his father had a farm of about 200 acres. With his brother, Matthew, George Culley was sent by his father in *c.* 1760 to study farming at Dishley in Leicestershire under Robert Bakewell, who was already achieving great fame for his breeding improvements to Leicestershire sheep and Longhorn cattle. At Dishley the Culley brothers acquired a life-long interest in improved farming methods and stock-breeding; they also met many of the foremost English agriculturalists. In 1767 they took the lease of a farm at Fenton near Wooler in Northumberland and began to put their new ideas into practice. They introduced new crops, new methods of cultivation and new rotations, drained marshy lands, erected new buildings and by selective breeding produced improved shorthorn cattle, horses and pigs and most notably, crossed the Dishley breed of New Leicester sheep with the native Teeswater breed to produce a new and widely-acclaimed Border Leicester or Culley sheep. They also became well-known for their improved strains of turnip seed and red wheat. In 1786 George Culley published *Observations on Livestock*, based on his own experience as a breeder. This became very popular; four editions were produced and it became popularly known as 'Culley on Livestock'. As a leading experimenter and practical farmer, George Culley was consulted by the new Board of Agriculture, which was established in 1793, and together with John Bailey, estate agent to Lord Tankerville of Chillingham Castle, Northumberland, he produced *General Views of the Agriculture of the Counties of Northumberland, Cumberland and Westmorland* in 1805.

The Culleys' farming innovations prospered greatly; they were able to lease more and more land in Northumberland and by the end of the eighteenth century they had become substantial landed proprietors. In 1795 they purchased an estate at Akeld on the river Glen near Wooler for £24,000, and in 1801 bought Eastington Grange for £13,000. Finally, in 1807 they acquired a large estate at Fowberry Tower for £45,000. It was to Fowberry Tower that George Culley moved to live with his son, Matthew, and it was there that he died in 1813.[16]

The Culleys maintained contact with farmers and stock breeders throughout England and Scotland, and the letters which they received from George Boswell and which are printed here, are part of that extensive correspondence, now preserved in the Northumberland Record Office.[17] The relationship with George Boswell came about because in 1787 George Culley asked his former mentor, Robert Bakewell, to suggest someone who could advise him on creating water meadows on his land in order to provide early grass for sheep and cattle. Bakewell recommended George Boswell. The initial contact was followed by Culley sending a man (Henry or Harry Rutherford) to Puddletown to learn the techniques of making and maintaining water meadows, the Culley brothers became enthusiastic advocates of the advantages of water meadows and the correspondence with George Boswell was maintained with occasional letters over the next twenty years. The surviving examples of this correspondence, with their interesting details about Dorset agriculture, are printed here. Culley's letters to George Boswell have not survived.

George Boswell's life was not without difficulties and disappointments. In his letters to the Reverend Dr Philip Lloyd during the early 1770s he refers on two occasions to the fact that his wife, Sarah, had given birth to still-born children, and the Puddletown parish registers list other children who died in infancy, including Sarah in 1763 (another daughter named Sarah was born in 1768) and Robert, who died in 1775 aged 11, and for whom the register lists 'decay' as the cause of death. His wife Sarah died in 1777 at the age of 45, and in 1779 his eldest son, George, suffered from some severe mental disorder and had to be confined in the private lunatic asylum at Halstock.[18] In July 1790 he must have had a further breakdown, which is mentioned in Letter 10. No further reference to him has been found. His sister-in-law, who had managed his household after his wife's death, died in 1796 after a long illness, and his daughter Sarah's health also gave him cause for great anxiety. Although he was obviously an active and energetic farmer, Boswell also faced financial difficulties during the early 1790s, and in 1795 was extremely embarrassed at not being able to pay George Culley for seed which had been supplied to him. It may be that it was the feeling of shame over this episode that led to the gap of eight years in the letters between 1796 and 1804.

George Boswell died in August 1815 at the age of 80. In his will dated 28 June 1813 he described himself as 'George Boswell, gentleman, of Piddletown in the county of Dorset'. By this time his daughter, Sarah, had married and was the wife of William Noyle Esq.; she was left for her own sole use the income of £2,000 invested in stock at three per cent. To his three sisters, who were all widows, and to their children and to the children of his brother, Thomas, he also left small legacies. All the rest of his estate, including copyhold tenements called Arnolds and Sturmers and a leasehold estate in Puddletown called Duckshole, he left to his son, Samuel, who was also his executor.[19]

The letters reproduced here give interesting details of many aspects of Dorset farming. They include much on the management of water meadows, on the carefully-controlled grazing and folding of the sheep flocks, on the cultivation of grasses, turnips, buckwheat and new varieties of wheat and barley, on the introduction of drills, threshing machines and improved implements, and on the eagerness with which some farmers embraced the new ideas and adopted the new methods.

They reveal Boswell's many contacts with leading agriculturalists, and it is a tribute to the high regard in which he was held that, as shown in Letter 15, he was approached by the recently-formed Board of Agriculture to revise the *General View of the Agriculture of the County of Dorset* which had been produced for the Board by John Claridge in 1793. Boswell declined this invitation and the revision was eventually written by William Stevenson and published in 1812. The letters provide detailed information about livestock, arable farming, the complexity of the weights and measures used in the county, the use of coastal vessels for the transport of goods from London, the difficulties which tenant farmers faced from the 'gentlemen sportsmen', who damaged their crops and disturbed their livestock by hunting and shooting. Boswell also provides information on Dorset during the Napoleonic War, George III's visits to Weymouth and to farms in the neighbourhood, the concern about the condition of the poor and the fear of unrest, the fluctuation of prices and the difficulties faced by local farmers. Above all, the letters reveal the personality of Boswell himself, his enquiring mind and deep interest in all aspects of contemporary agriculture, and his eagerness to learn about and participate in the exciting new developments in farming methods.[20]

REFERENCES

1 J. Hutchins, *History and Antiquities of the County of Dorset*, 3rd. Ed., ii (1863), 615.

2 C.L.Sinclair Williams (ed.), *Puddletown: An Account of the Inhabitants of Piddletown Parish 1724*, Dorset Record Society, 11 (1988).

3 Dorset Record Office, PE/PUD: RE 3/1, Puddletown Parish Register. See also H.G. Chick, Transcript of Puddletown Parish Registers, 1937, copy in D.R.O.

4 C.L. Sinclair Williams, *op.cit.*, 22.

5 *Ibid.*, 51.

6 D.R.O. PE/PUD: RE 4/2.

7 J. Hutchins, *op. cit.*, 615.

8 D.R.O. PE/PUD/IN5/1/1 Puddletown Tithe Agreements and Correspondence 1695-1847.

9 Huntington Library, Stowe Collection, STG Correspondence Box 422. This letter was found in the Stowe Collection by Mr George Clarke and I am indebted to him for providing a transcript of it.

10 I am grateful to Mrs Jennifer Hawker for providing these references from the Puddletown Overseers' Accounts D.R.O. PE/PUD/OV1/21/5; PE/PUD/OV1/4.

11 D.R.O. D/PUD/M6-9 Records of the Earl of Orford's Puddletown estate *c.*1760-1796; D/FRA/E32 Leases on Frampton Estate 1798-1808; D/FRA/E69 Letters and Papers of James Frampton 1783-1807.

12 D.R.O. D/FRA/T9, 10 Frampton Estate, Waddock Farm Leases; D/FRA/E68 James Frampton's Memoranda Book, quoted by Joan Brocklebank, *Affpuddle* (Bournemouth, 1968), 82-3.

13 *Ibid.*

14 Joan Brocklebank, *op. cit.*, 83.

15 D.R.O. D/FRA/E69; Joan Brocklebank, *op. cit.*, 43; Puddletown Parish Records D.R.O. OV1/18.

16 D.J.Rowe, 'The Culleys, Northumberland Farmers 1767-1813', *Agricultural History Review*, XIX (1971), 156-174.

17 Northumberland Record Office, ZCU 12-27; D.R.O., Photocopy 415. The Dorset Record Society is grateful to the Northumberland County Record Office for permission to reproduce these letters.

18 The reference to the confinement of George Boswell junior at Halstock in 1779 was kindly provided by Mrs Jennifer Hawker from D.R.O. Quarter Sessions Papers – Private Lunatic Asylums.

19 Public Record Office, PROB 11/1571/417, George Boswell's will, proved 26 August 1815.

20 The letters all start 'Dear Sir', and end 'I am Dear Sir Your obliged obedient Servant George Boswell', 'Your obedient . . .' or variants, with 'Piddletown near Blandford', or 'Piddletown'. Except for two instances at the end of letters, these have been omitted from this edition.

THE LETTERS

An expression in your Letter which I received two days since, determined me to write to you immediately, (although I may not have time to answer all your Questions, which I shall with pleasure do on a future day), it is this, *'I believe I should have been able to have persuaded my Brother to have gone up and seen the mode of Watering, but etc'.* I shall be extreemly glad [to see] either or both of you, as well on our Friend Mr Bakewell's account as from the congeniality of *our* Sentiments. I am engaged to be near Thetford in Norfolk at Mr Colhoun's at Wretham, the first week in June, upon the subject of Watering Meadows – He is an entire Stranger to me. He is a Member of Parliament and possessed of a considerable Estate there, and an intimate acquaintance of Lord Orford's who is Lord of the Manor of Piddletown where I hold a small Estate under him, and his Lordship referred him to me – as our Friend did you. Norfolk being my Native County and some other reasons made me think it proper to offer my opinion upon the spot, and afterwards to have a man down from thence or to send from hence etc., etc. As it is probable I may never be as far Northward as that again for many years, if ever, if you or your Brother could make it convenient to meet me at Thetford, much information may be obtained on both sides, viva voce, which cannot be communicated by Paper. Some things you have mentioned have raised my curiosity, particularly *Spring Wheat* and *Early Oats*, but no more of that for the present. Having so far settled about the Man's coming, little more need be said about him. His Speech and Dialect will be no objection. I have the pleasure to say this part of the Kingdom amongst the lower Class are not deficient in civility to Strangers, and I shall with great readiness take him under my Protection and shew him some degree of attention.

You ask if you direct properly; I am a Tradesman and a Farmer, therefore entitled to no other direction. When our Friend was here, I had a considerable employ under a Nobleman in this Neighbourhood, but his sentiments and mine were so totally opposite. The Tyranny of an overbearing Lord and the honest freedom of a British Spirit could not long be in unison, so I returned to the Plow etc., again. You ask *'What kind of Stock etc.?'* When Meadows have been watered some Years and are become tolerably firm and the Herbage pretty fine, We spring Feed those Meadows with Sheep either Wethers or Ewes and Lambs, from about Lady Day to near May Day – but never with Sheep in the Autumn, the Autumnal Grass whether watered or not will infallibly bring the rot upon Sheep, therefore that Grass is fed by Dairy Cows or Fatting Beasts or Young Stock of either kind whether Beasts or Horses, till the Watering season begins which is in some meads the beginning of October in others not till near Christmas – Water Meadows are never mowed with success twice, the damps arising from the low grounds in the Autumn prevent the Grass from drying well, and even when cut and carryed upon the highland, it consist only of a spiral blade which when withered, feels in the hand as sort as Wool, and though excellent in *smell,* in quality not so good as straw.

I must caution you, how you put your sheep into New Made meadows. Never attempt it (I mean in the Spring) till they have been well watered, and even then put at the first trial a few only. Keep them there till you[r] Meads are lain up, suppose the end of April, then take them out and let them go with your Flock, and in September kill one of them and examine the Liver – if not sound, forward the others for the Butcher as fast as possible, for as soon as the frosts come on they will pine away and be worth nothing – if that so killed is sound – let them run with your Flock all the Winter, and if they retain their strength etc., they are sound and the Meadows safe. I must inform you that if a water Meadow, ever so good, should not be watered all the winter – the Spring Grass will certainly rot Sheep – no danger to any other kind of stock.

128

We don't like to Feed our Water Meadows, which are intended to be mown for Hay, in the Spring with Heavy cattle because they tread the works so much, that the water can't be carryed regularly over the Land. But in the fall of the Year, or Autumn we don't mind how much they are trodden. You will find the Spring feed excellent for your Young Calves, but not in the Autumn. We seldom take in Stock when it's done in the Autumn a person take a Meadow at 8s od, 10s od or 12s od per acre for the *aftermath*, that is after the Hay is carryed off the meadows are eaten out by the owners Stock for a few days – then the water is put over the land as so[o]n and as often as it can be done, and by the begin[ning] of [Sep]tember or sooner there will be a good head of Grass which is *let* till watering time. In the Spring the Butchers will often put their fat sheep into water Meadows at 4d per head per week – Ewes and Lambs sometimes are taken in at 6d per head per *Couple* i.e. *Ewe and Lamb*, prices vary exceedingly, besides this Country is so full of Sheep that every person almost feed their Grass with their own Stock. But as I hope to meet you, no more of this at present. Pray *remember*, once for all. We are now upon an equal Footing, the Questions etc., etc. I may ask of you, will be reciprocal, and therefore postage of Letters must not be paid.

Letter 2   25 March 1787

I am much obliged to my friend Mr Bakewell for his good opinion, the best return I can make at this distance is to shew my readiness in answering yours.

The method you propose, I think the most judicious, for besides the Expense of a Man's going from hence, his extra wages, etc., etc. by his self consequence, and *acquired* importance it is more than possible he might withhold much useful instruction etc. The great distance you are at from us, you are well aware of, and consequently the expense of a man's travelling, time, etc. If that is removed, the method I shall submit to you is: to fix upon an healthy, robust Man, who has been *used to labor*, can write intelligibly, at least to himself, sober, and about thirty years of age. If he understood a little of drawing it would be a great advantage to him, but no insuperable objection without; It is absolutely necessary for him to *be a Laborer* and to be both willing and able to go through the manual part of the work in all weather, as the Watermen do here.

If he is clever and ingenious, and the work you shall want him to undertake on his return not extremely difficult or extensively large, I should think one Season might be sufficient to qualify him, though I apprehend you will think with me, it is but a short time, and that another season would not be too long.

The time He should be here by, will be the beginning of October and remain to the end of April – that is the full season for watering Meadows with us, that are to be mown, but when there is a capital peice of *new work* going on, then, (as you will easily conceive) the work is continued the whole year through.

I have talked with my Waterman (who does all my regular work by the *great*, that is at a certain price per acre, and not by the Day) and told him, it was possible a person might come from a great distance to learn to water meadows for about seven months, four of which at least be with him, perhaps more, but I reserved a power of sending him to work two or three months of the time elsewhere if I thought it necessary, to work under different men and see different kinds of work, for *New work* is not always going on, in a country where watering meadows have been used perhaps more than two hundred years.

A very few years since I watered above twenty Acres which were never watered before, that will be a model for him, if well attended to supposing no other work should be on hand.

My waterman said He would engage to give him 5s per week for his work (if He worked as he did) for the four Months or any longer time, only He expected some little gratuity for his instruction. You may possibly think the *Wages low*, but He said he certainly would not take a person to shew him the nature of the employment, without an advantage arising from it. Laborers are hired here by the Watermen for from 6s to 7s per week, and after having worked for years are not at all qualified to perform the work by themselves unless it is on the little Farms where the Masters work with them etc.

You will understand the Man is to find himself in diet and Lodging etc. The only things that will be wanting here are *a pair of Water Boots and a spade*. These must be had here – his Cloathing should be coarse and strong with a strong coarse great Coat, for they are out in all sorts of weather. Perhaps, you will think, I have formed so many requisites etc., etc., that you give up the Idea at once – if so, we are just where we were before. But if otherwise there is one thing more, that I must recommend to you, which is to let *him* go and work with a *Carpenter* this summer. He will be able then not only to *see* the manner the Carpenter's work is done, in making and setting the Wares, but will be able to assist your Carpenter on his return, a matter of more consequence than you can be aware of, but I will drop this subject lest I disgust you.

I thank you for your information relative to the prices of Corn etc., local expressions are so familiar to us, that we adopt them with the Idea that they are known everywhere. You say fat Cattle are sold from 4s 4d to 4s 9d *sink offal,* that is I suppose the Quarters are sold at that price, and the Hide, Tallow, etc. are included in the bargain. If I am not right I will thank you to set me so, the 4s 4d etc. is per Stone of 14lb. Fat sheep at 5d per lb. means the same. Horses with us are enormously dear, beyond anything ever known. Fat Cattle bear a price beyond all bounds. A good fat Heifer which will weigh from 30 to 40 Score pounds (20 lbs to the score) quarters *only* reckoned will now sell from 8s to 9s per score, Hide etc. included in the bargain. Lean, barren Cattle are very dear full £2 upon £6 advanced. Fat Sheep are very dear I sold 80 out of 100 for 30s each supp[ose]d to weigh about 19 to 20 lbs. per Quarter – the rema[ining] score I sold half at 4¾d per lb the other at 4½d per lb by wei[ght]. Fat Pigs, owing to the great plenty of potatoes to feed them sell at 7s per score lb weighed after they are dead, and lean ones for the same reason immoderately dear, 6s upon 10s advanced. Best raw Milk Cheese from 5d to 6d lb, hard ordinary Cheese made after Butter 3d per lb – and Butter from 10d to 12d per lb of 18 oz – Wheat 5s 6d to 6s per Bushell from 68 to 75 lb weight, Barley for seed from 24s to 26s per Quarter of 8 Bushels, Oats from 19s to 21s per Quarter. The Bushel here is more than 9 Gallons. Fleece Wool from 23s to 25s per Weight of 31 lb. Lambs wool 24s to 25s ditto. The two last Summers from their extream drought nearly ruined half our Farmers. We have had a very fine winter and remarkable Spring – but a heavy rain is now set in. Oats and Peas have been some sown these 3 weeks and Barley more than a week. Your answer relative to the Man will in a reasonable time be expected for the Waterman to make his arrangement. Any further information will when requested be at your service.

Letter 3   9 September 1787

Having just finished Harvest I mean to discharge the obligations I am under to my Friends for their repeated favors, amongst the first You Sir have an undoubted claim.

I was in Norfolk when You[r] kind Letter came into Dorsetshire. I am quite satisfied We have hit upon the only method to insure success, that of sending a man hither; since my return I've been looking out for one to send for a month or two only to Mr Colhoun, the Gentleman whose land I went to see, whether it would be likely to succeed; there was no objection that I could observe, except the want of *Water*, which He was so sanguine of obtaining in any Quantity, that removed all my objections and told him I hoped to send him a person to lay out the land properly etc., the rest must be his concern, and so I left him – satisfied in my own mind, that the Works might be cut, and in the Winter partially overflowed; I don't dispair of getting a man as the distance is not great and the time short. My Waterman will be ready for your man when He come, which I wish may be about Old Michaelmas, the autumnal Rains having generally set in by that time. Our Wheat Crops are very light and thin. The Barley in general thin upon the Ground and hurt a good deal by the Clovers which are remarkably forward this season Oats a middling Crop. Wheat is got up to near 8s od per Bushel and old Oats to 24s and 25s per Quarter, our Bushel more than 9 Gallons. The *Pur* Lambs, He Lambs cut, have sold and are still selling well this year from 11s to 16s per head – the females are seldom sold, but by exchange of Farms or a Farmer's Quiting business; being kept for breeding. The Old Ewes will sell well this season the buyers are come down into the Country already, but the Ewes are backward with Lamb this season in general. I fancy your's never Lamb as soon as Ours in this Country. (I don't keep an Ewe Flock). We have many Lambs fall by the 12th October and a great

132

many by the beginning of November and the principal Stock of Old Ewes have finished by about Old Christmas. Wool sells well 24s to 25s and some 26s per Weight 31 lb., about 9 fleeces to a weight.

Much has been said relative to the advantage of change of seed, from a distance. Although I am not convinced by any argument I've heard; yet wishing not to dissent from every received opinion I subscribe to it, should your Ideas correspond with mine and be inclined to exchange twenty or forty Bushels of Barley at the market price of each Country, delivered at Each sea port. Our Barley is remarkable for its Quality for malt, and may be had clean. If any other Grain should be thought of, I shall be happy to join in such a scheme. For instance have You Tares or vetches that will bear the Winter; we have none that are good, but we have remarkable fine *Spring Vetches* sown about Lady day. If it is do not take place we may talk about it *[sic]*. When I was in Norfolk that part was remarkably poor Land, they sowed Rye with success – I am going to adopt the plan having a great deal too much poor land for my profit. Have You any Grain particularly adapted to such soil? I can say nothing relative to our stock, indeed we are at too great distance to benefit by either, but we are very inattentive to Stock of any kind except some [far]mers in respect of their Ewe Flock – and even there [they are] well contented if they don't decline in Quality, hav[ing] scarce a wish to improve (I mean with any Expense). Mr Bakewell will never make converts of our Farmers for He is esteemed here the best Farmer, who gets the most Money, and which is generally done where there are least outgoings supposing a farm is rented *accordingly*, you will not subscribe to that Idea; at least I can be assured you do not Farm under such an Idea, nor can I though I see instances of the fact often. It is true their Crops are not good, but they are at *no expense*. My paper is done. I shall always esteem it a favor to hear from you.

Letter 4   27 December 1788

It is an observation not more trite than true that when we defer that till tomorrow, which we could have done today, we are just were we where *[sic]* when the morrow comes, it may be applicable to me, in answering yours, though that is not the only reason. Your late loss (which I sincerely condole with you for) call to full recollection the happy days I spent for 17 years though the last of them is now Eleven years since. The impression is strong upon my mind now, but not the appearance; Time and business are the best comforters, for the one employ[s] the mind, the other deadens the feelings. Another strong reason was, I knew not how to answer it – if there be a pleasure in doing a friendly action (which is at the same time a duty), there is also real pain in being told of it, particularly when the gratitude expressed far outweighs the service done.

I am glad to hear Harry got home safe and well. I flatter myself, He will be able to shew you by a *little*, which is the scale I wish him *first* to go upon, what improvements may be made upon a *larger* scale. You will remember the work is expensive, but once done, afterwards it's very moderate, one great thing is, the produce goes to assist in the manuring other Lands, whilst Watered Meadows want none brought on them, except Water: you say I bear *too hard upon* you etc. to remain with that impression upon me, may be construed into *Vanity* or *Pride*, both which, if I know myself, I abhor. If you have any opportunity by the Coal Vessels to *London* or *Poole* the latter best, the former more certain, to send me a Sack of each of your own Country Oats. Harry knows my soil at *Waddock*, and will be best able to judge which will answer there or at *Park:* any Grain adapted to your Northern Climate, would improve here, how far the Southern grain will agree in the Northern I cannot judge but some of our Barley is at your service. Corn was a very indifferent Crop with us last year, owing to the dry Season, Wheat now sells from 6s 6d to 6s 9d. Barley 2s 9d. Oats 2s 3d. Peas White 6s per Bushel. 9 Gallons or more – the amazing Crops of Apples in the neighbouring Counties has made the Barley a dull commodity at Market. One parish w[h]ere my Son was the other day will make between 7 and *8000* Hogsheads of Cyder, another parish whose real annual Value is not £1000 per annum will make near 3000 Hogsheads. We make but little in this part of our County. Our Fairs concluded this year much against the Jobbers – Sheep sunk 3 and 4s per head at the latter end – though even then great prices. The Ewes that were forward – that is, were beginning to yean sold from 26s to 32s per head at Weyhill fair.

Horned Cattle sold Dear, Horses and Colts very much so. Pigs sink in value much lean; fat yeild 5s 6d to 6s per score lbs.

We have had such a series of dry cold Weather as was never known before – no rain except part of one Day since the beginning of October, many parishes not a drop of water, and are forced to drive their Cattle two and three Miles for Water.

Tell Harry, that I've not had one acre of land watered this season – that the Diary [sic] Cows go every day into the water meadows for the rough Grass. I had built Cow Houses when He was here for 26 Cows at Park. I have since the beginning of November, began building and finished at Waddock Cow houses for 43 Cows and at Piddletown for 17 more, and very astonishing to say it, the workmen were not hindered by the Weather half a day the whole time. Cow houses are new things here. The Bishop of Llandaff has written twice to me upon the subject of Watering Lands near Winander Meer. He made use of Your Name, I am much obliged to you for your good opinion, but I don't think myself at all equal to i[t] the Bishop's Public Character has a full claim upon e[very] Individual – my mile will be readily thrown in; but you should consider that when our friends are too flattering and say too much, they disappoint the third person, because they are taught to expect *much* but meet with *little*. The Bishop mentioned his Steward Mr Benson as capable of superintending, if he had an opportunity of seeing any work – I have made him an offer of *my House* and my assistance, if He will permit him to favor me with his Company. The Coach in one Day brings him very near my House from London – Suppose you accompanied him – a Peice of Beef and Pudding with some Dorsetshire Beer, and two dry warm beds will be, with a hearty welcome, the reception you will find from, Dear Sir, Your most obedient Humble Servant, George Boswell.

The folks whom Harry knew when here, one and all are pretty well, and those belonging to me whom he remembers return their good wishes. The little Mead at Waddock west from the House, which I underground drained is now much improved by watering, although before it was of little service owing to the Cold weeping springs which ran under the surface and poached the soil – that is very expensive work and cannot be recommended unless where the land is an object – The frost has hindered us plowing ever since the 24th November and has continued without scarce any remission. No Snow at all. This Evening a little falls – December 29th.

Letter 5 24 January 1789

If you thought the Bishop of Llandaf was the spur to my answering Your last favor, what will you think the motive of this is Self Interest. Exactly so, and when we can pleasure friends at the same time there is no harm in it. I thank you for your obliging favor, and will accept your offer of sending me a Sack of each of the *four* sort of Oats you mentioned. If you have sacks large enough to hold 5 Bushels I should wish to have them, and if at the same time you sent me a Sack of your best eating potatoes I will thank you, if made up of two sorts we can seperate them. You will be pleased to direct them for me at Piddletown to be put on board a Pool Vessel from Cottons or Chamberlains Wharf, London. Pray let me know when they are sent, the sooner the better as we sow early. I will take care to send some *clean* Barley, for Seed, it will not be good in bulk this year, our Summer was so dry, but will be fit for Seed. The long frost has at length left us, no Snow till two days before the Thaw, which was accompanied with a great deal of snow and rain, which melting as it fell, and the ground so hard, occasioned a very great flood and did considerable damage to the Wares upon the Rivers. Pray tell Harry Hurst bridge is gone and a great part of Mr Frampton's bridge at Moreton. I beleive I forgot to caution him against keeping the Hatches to the Wares upon any considerable Stream shut, when the frost appear to set in hard, for if they are frozen in, when the flood first come down it will be dangerous of blowing up the Wares.

I have been some time endeavoring to invent a Machine for Drilling all sorts of Corn and Seeds, upon a different principle from any I have heard of, for I have seen none except Drawings. I constructed one late last Wheat season and it seemed to answer very well, that is the principle upon which it acts, for amendments it will want – to all that I had heard of one objection was strong with me, *the price* and its complex form; both which I think I shall

136

get over. One difficulty I met with and which I was not aware of, the delivering more grain out going down the hills than when going up – that I think I have now got over. I shall be able to make the Drills and sow my Turnip seed, both in those drills and broad cast at the same time, that when the Turnips come to hoe, if there should be a failure in any part of the drills, the others (if any) may be left to supply the deficiency. I shall be able to drill at any distance, but my principal plan is for Corn and I have it in contemplation, and I think with certainty, to sow Barley or Oats in Drills and the broad Clover broad cast at the same time and with the same machine. I would not wish to be thought too sanguine like most scheemers of their own inventions and perhaps I shall not bring it to perfection, yet I have no doubt of the possibility and practicability. Corn remains in the price nearly *[damaged manuscript]* when I wrote last. Barley is buying up by the [mer]chants (upon Speculation I believe principally) to be carried Coast ways where it will pay them best. Their price delivered at the ports 22s per Quarter, which is about 20s Winchester. Grass seeds it is thought will not be dear this season, about 2s per Bushel, 36 Quarts, we sow from 3 to 4 Bushels per acre.

We are now busy in *settling* the water in our Meadow (Harry understands me) had the Water not come soon we dare not have Spring fed any of the Water Meads this Spring, for fear of the *rot*.

Letter 6   6 April 1789

I have the pleasure of informing you that a few days since your Oats and Potatoes arrived safe, the latter I have not opened – but the Oats I've this day finished drilling in about Six Acres of Ground not so good as I could wish, but perfectly clean – I drilled them in a day and half in rows a foot asunder – rather thicker in the drills than my own judgement approved of, but the prejudices of the Country are so great – that I rather chose in some degree to coincide with them, than absolutely to run counter to them all.

Each of the kinds are vastly superior to any we grow here but the Poland are infinitely beyond any ever seen in this Country. The Angus are my favorite sort. I bought Six Quarters of Poland Oats imported at Weymouth from the Continent, though thought here very good yet when compared with your's, they fell very short indeed – those cost me at Weymouth 22s per Quarter or 2s 9d per Bushel of 36 Quarts. I sent sometime since 3 Sacks of Barley agreeable to your directions, which I hope will come soon to hand – Barley is getting up with us, I sold saturday for 24s per Quarter for Seed – Grass seeds are very cheap, tolerable good ray Grass or Darnel for 12s to 14s per Quarter – I sold all mine but it had some Black seed or Nonesuch mixt at 2s per Bushel – Broad Clover is now selling at 4d, 4½d and 5d per lb. of 16 oz.

My drill plow will I think answer my expectation, Oats and peas I've drilled a foot asunder the rows, 5 Rows at once, drawn by two Horses, but it's hard work for them about 5 or 6 computed Acres per Day – I think if I add a third Horse I can drill 7 Acres per day. You I suppose have seen *Cooks*; I have not, pray give me your opinion upon it. I took your hint respecting the Bishop of Llandaff, and have heard no more of him upon the subject of Water meadows. The Member for Bedford made me a little shy in offerring services etc., etc. as soon as their own turns are served little folks are forgotten. Believe my good Sir there is more reall sincerity and friendship, (at least I have experienced it) in the middle line of Life than in either of the extreems. Have you seen a new treatise upon Water Meadows by a Clergyman in Gloucestershire, if not I dare say you will. People in their zeal for a Cause they've espoused generally think they can never point out the advantages too strongly – but by over doing it they often hurt it, such you will find to be the case there. I had much sooner know *how to make the most* of my land, than to be told *how much* can be made of it. If I know the former, the latter will follow of course, and *the farmer* not hurt.

Your County I saw supported the *Foxites*. We are all *Pittites* here. I have only to say, that I never knew a *wrong* cause so *well* supported. We are all happy at the lucky turn. Your cause have many able and good men with it, but they are united with a number of desperate and needy harpies. The Prince is a generous, lively and gallant *Young Man*, the King is an amiable and virtuous *old one*. So much for Politics. We have had a very long winter, the Turnips mostly destroyed and the spring so backward and the water through the winter so scarce that we have little Grass in our Water Meadows – Stock was never known[n] so poor and bad. I was the other day at the Sale of Stock sold under an execution for Rent, the Couples (Ewes and Lambs) sold from 23s to 25s per Couple, Wethers from 17s to 19s per head – worth more money last Michaelmas, and the (Hogs) last year's Lambs sold from 10s 9d to 15s worth more at Michaelmas and have been out the Winter at Keeping at 5s per head expense, but I never saw anything so villainously kept in my life, many of them must die. Hay is very scarce at many places around us. I have not sold my fat wethers yet, I believe I shall fatten no more, but sell them off getting them pretty well in order, about Michaelmas – the quantity of food they eat is enormous – they are good, but I have given them some Corn, Buckwheat and Rye mixt with cut Hay in troughs, about of 2 pecks ½ per Day to 100, and has been of vast service. The Butchers come to see them tomorrow – three score out of an hundred and ten for the first draught – shall I get Two Guineas per head for them? I own I expect it – they will weigh 22 lbs per Quarter average.

I mention to you the quality and quantity of Corn given to them as an hint to you, Rye alone with cut Hay I think best. The Butchers have been with me – We could not agree in a price per head they offerred 40s – I wanted 42s, we agreed for 5½d per lb. the quarters – mutton is not worth so much as last year – tho[ugh] there never were less *thorough fat*, but half fat is very p*[space]*. Monday I shall begin drilling Barley in drills 8 Inches asunder 7 ranks at a time. I think we have got over all our difficulties and our Machine seems nearly compleat.

Pray present my best respect to Your Brothers, and tell Harry all his acquaintance here are very well – for these two last days the Water meadows begin to grow – they were scar[c]ely ever known so backward before. I've just received a Letter that your Barley is gone in a Poole Coaster for London, I hope you will receive it in time. *[Letter dated 10 April 1789 at its conclusion]*

Letter 7   25 October 1789

The business of the summer being nearly over, it becomes our duty to look round us, and see how we have spent it, to endeavor to correct what has been omitted, or neglected. In the latter list, your name stands conspicuous, and stares me in the face, with a silent, but severe reproach; but as I mean to own my fault, I fancy I can see the smile of benignity, and forgiveness on your face; thus encouraged, I begin with telling you that a wet cold summer (which produced but a short crop of Hay in *our Water meadows*, and but a middling time for taking it) we approached Harvest, with no promising prospect a wet week in the midst of our Wheat Harvest (which, as we mostly mow was great part in swarth received a good deal of damage). We find the produce very small this year, average I believe not more than 14 Bushels, nine Gallons, perhaps *certainly* not so much – and sow three. Barley on the light lands good – on the heavy, very thin corn. Oats in general good. This bring me to acknowledge the favor of your's, each sort seperate in the same field, but in different land – a fine tilth after Rye sown with my drill plough. Oats are not a good grain to drill, for the form of the grain contribute much to obstruct the regular discharge – many parts were injured by the worm etc. yet upon the whole I had a very good crop and are all saved for Seed, though I shall be tempted to have more by and bye. They were all sown within one day of each other yet the Poland were cut first 26th August, the Dutch not quite ripe cut the same day – the other two were not cut till 15th September – and then the[y] were not ripe. The Angus and Poland seem to be the sorts best adapted for us – the stalks of many of the Angus were the largest ever seen here and the Poland Oats the best that ever were seen in this Country. I had some Poland Oats that were imported to Weymouth, but I believe they were the same as you called the Dutch. The produce of each of your's and of those are saved seperate not for the sake of ascertaining the Quantity but the quality. I last night carried my Broad Clover for seed, it is expected to be dear this next season, but I don't think so. You thought that Broad Clover seed might be bought here to advantage – the change of all those Seeds is considered as one, and the freight a triffle – If I can buy you any hereafter, you may command me without Ceremony. You must in that case determine the price you will give – if it run dearer here, I shall buy none; if cheaper the better – I shall certainly wish to have 20 Sacks of Oats *in time*, of the Poland and Angus.

Wheat before Harvest got up very dear 9s 6d and 10s per Bushel 9 Gallons, once for all, when I mention Bushel I shall always mean 9 Gallons unless other ways expressed – What we call

9 Gallons [?is] at least a quart above 9 Gallons and some Farmers two. – Since Harvest for seed worth more than 8s 6d – and the millars give 8s there is a good deal of grown Wheat this year with us – I have just finished Wheat and Rye sowing, many with us have not near done yet – and in Somersetshire they have but just begun – Barley sells from 22s to 24s per Quarter. The Last Lamb and Sheep Fairs fell considerably – The low land people have not been able to buy any – and almost all they have are rotten – one Farmer bought lately in Somersetshire 1300 Sheep that had been worth from 20s to 28s per head, for from 8s to 10s per Sheep and drove them to Portsmouth for the Convicts etc. – and got a great deal of money by those that came to Market, but a great many dropt by the way. Pigs this season have been remarkably cheap and have sold fat for 4s, 4s 6d and 5s per Score 20 lbs. Beef now sells in the market for 3½d and 4d per lb. Mutton 4d and 4½d – Butter now sells for 8d, 8½d and 9d per lb. of 18ozs, perhaps the best in England – Seed Rye have sold for 4s 6d and more per Bushel. Wool sells well 10d per lb. – about 4lbs per fleece – smaller fleeces will yield more.

Our markets of all kinds are yet hardly established – so that the prices are not at all fixed yet. I have had repeated applications for another Edition of the Treatise upon Watering Meads, (the former being quite out of print) – A Clergyman of Gloucestershire having lately published upon the subject – so very superficially and in a great many instances very erroneously – I have been obliged to enter the lists and to animadvert upon his publication, much against my inclination – He has stated some facts directly opposite to what I had advanced so strongly – that a Gentleman Mr Keld of Beverley in Yorkshire, came to my house, on purpose from London to ascertain the fact (viz. that water as clear as Chrystal issuing from springs near, has made as great improvement upon the land as can possibly be conceived) – I carried him to the meadows and convinced him. To him, in a New Edition which will come out this Winter, and also to *your Henry*, I have referred Gentlemen in the North of England who may wish to have the matter cleared up; for their satisfaction: Henry knows the meadows which I mean; they are those just above Piddletown and also quite down to Park farm where he worked – the water that supply those meadows rise but a little way above us and is as

141

clear and pure as can be conceived. – How does your Threshing machine succeed – is it complicated and expensive? I have kept my Barley that was drilled by itself to know the difference of the produce compared with Broadcast – I shall next Spring I believe drill a good deal. I have s[een] a new machine for slicing Turnips for beasts – it cuts them I think too small – There is coming into use here, a machine for bruising Horsemeat – such as Oats, Barley and Beans. I have not seen any used yet – but it is spoken highly of, for bruising Oats – to prevent Horses swallowing them whole.

October 31st. A Dry Northerly Wind – the Air seems now settled and believe we shall soon have frosty nights and clear days. Just finished wheat sowing. Corn this day's market nearly as the last. Barley seems rather advancing, and must do so I think whilst we are under the London markets. Last summer I passed over a farm where there were near 2000 Sheep – I was informed (for certain) that there was not one Acre of Arable or Meadow upon the farm, the whole consisting of *Down*land. – I leave you to guesss how he keeps them in the Winter! Pray tell Henry *those* He knew here are all pretty well and much in the same state. George Hoar often ask for him.

Letter 8   23 December [17]89

It's with the justest sense of the obligation I acknowledge the favor of your last, with Mr Bailey's drawing of the Threshing machine. I should think myself peculiarly fortunate in an opportunity of thanking him personally, in the meantime let me request of you to do that friendly office, it will stamp some value on it, by your manner of doing it, with him I am sure. I am perfectly master of it from the drawing, excepting one part, viz. the *motion of the Rollers.* I presume the upper pulley round which the rope runs is united by a spindle to the under Roller. How has the upper Roller motion communicated to it? and what keeps the two Rollers at equal distances from each other? I suppose that the inclined frame in which the rollers are fixed and to which the weight is fixed is alike at each end of the roller. I observe it is said that the weight should not be so heavy, but that the man may easily draw back the rollers with his hand – it appears to me that as the

Rollers are kept by the weight beyond their perpendicular towards the switchers, the Rope which runs round the pulley would not permit the rollers to be drawn back by the hand. I have no doubt but it does, and therefore I do not clearly comprehend that part. Pray request Mr Bailey at his leisure to furnish me an explanation, and it will save me some calculation if you tell me the *Diameter* of the great Wheel which has 160 Cogs on it. Don't think me impertinent! I know it is said of old, *No Shoemaker beyond his last.* Yet I can't help thinking, the ingenious inventor, might easily fit to it a machine for cutting *straw* or *Hay* for Cattle into chaffs I dare say you practise that method by a hand engine. I could wish Mr Bailey to hint it to him, if *he* have ever seen an Engine to cut Tobacco that go by water, he will take the hint immediately – if not – I think if the Wheel to which the Switchers are placed were made with a given number of breaks in them, like the rack wheel of a Clock – so as to raise a large Knife up to a given height – to which knife a given weight was added to increase its velocity when it is discharged from the Wheel, and so to fall in a grove just before the rollers. I think it would do a great deal of work in a day, fed by a man in the same manner as the Threshing machine – The mechanic who so ingeniously invented the machine, will from the hint instantly see whether it is practicable. If you have not used the cut straw or hay instead of Chaff yet, you will find it a valuable acquisition. Last year our Turnips were by the severe weather very much rotted. I had more than a Hundred Sheep then fatting and could not get them forward – I made some troughs (ten) into each of which I put once a day half a peck of Buck and Rye mixt with a quantity of Hay cut fine with the Engine, it is amazing how they *proved* with it, the Expense of the Corn and *cutting* the Hay not three farthings per Sheep per day. I sold them for 40s per head 3 score out of a 100. Some of our people turned their's back to the flock again – others who bought them at Michaelmas at 25s sold them at Lady Day for less money.

You ask me if I were ever at Dishley, No – That it would be the greatest treat to me certainly Yes. But I cannot go this Season that is equally true. January or February are surely not propitious Months for an Agriculturean expedition. April and May are much better, and not exceeding busy Months for the Farmer – Every person franked from Northumberland with *G C* will bring his welcome with him. The operative part of our Husbandry may be soon learned, except what the vicissitude of Seasons brings with it. I am certain he will hold our management of Cattle cheap and but little to be learned from it, except our *forward Ewes* – many Lambs in allready. Red Clover Seed is not so much as thought of yet with us – We hardly know the price of it here before Lady [Da]y. If you know the price hereafter that you can buy it for [and] will let me know, and if our prices should be considerable and will commission me, I will do my best. Do you buy ra[y] Grass seed in your parts. We use a great deal here, it is not remarkably good or bad with us. If Mr R[ichmond] come in time He will see our kind of Seeds etc. here and report his opinion.

Not a word about the Dorsetshire Barley! a bad sign – You ask me what quantity of Oats I should wish – 10 *full* sacks of Poland and the same of Angus. Wheat now sells at 8s 6d and old wheat at 9s per Bushel, Barley 3s 3d and 3s 6d – Oats 2s 3d to 2s 6d per Bushel more than 9 Gallons – if you should wish to try any more Barley mention it in time. Seed Barley will be near 4s it is suppos[e]d 9 Gallons – Beef alive advanced last week from 6s to 7s per Score – Mutton will it is thought be uncommonly dear – in market Beef sells for 3½d – Mutton 4d – Veal 4d Pork 3½d – best Compliments etc. etc. etc.

Christmas Compliments and New Year's wishes attend you both. Why do you grow Buck: what sort do reckon best and how apply it.

Since I wrote the above I have reason to think that by the *same motion* and at the *same time* a Chaff cutter might be affixed opposite the Corn rollers – with another pair of Rollers and the Knife lifted by the Switchers, as they rise they will raise knife etc. pray let Mr Bailey think of that.

Letter 9   14 January 1790

Pray will a Thresh[ing] Machine swindle flax. Your anxiety for Mr Richmond induces me to address you immediately, although he have written some time since, but for fear his Letter should have miscarried, I have the pleasure to say he is very well, and don't seem to regret his Journey or time. We shall do what we can to make the one agreable and the other I hope not altogether unprofitable, for the rest I refer you to himself. I hasten to acknowledge my particular obligations to Mr Bailey, to whom I beg you will make my best Compliments and grateful thanks for his obliging attention and trouble, which I shall retain a proper sense of. He has I think made everything perfectly intelligible and I am not without hopes that it will be attempted here by and by. I will thank you to ask him, 'What distance the under Roller moves from the inclined board on which the unthreshed Corn is laid and by what means the unthreshed Corn is prevented from slipping betwixt the roller and the concave board.' I am aware the under roller turns the opposite way, but Query 'Whether the Corn do not sometimes muffle the roller, or at least what prevents it?' Is one or two Horses requisite to work it.

I observed Mr Bailey said in enumerating the advantages, bruising Flax Seed for Calves, Pray how do you use flax seed for Calves! I will thank you to tell me, I want to rear Young Calves without any skimmed milk or with the least possible quantity, whether it is practised with you and how? Red Clover seed price is never known with us till very near the sowing Season, by that you will find our Growth is little more than the Consumption. Direct for me per a Poole Ship from Chamberlain's Wharf London. Buck wheat is sown for many purposes for plowing in, as an excellent manure for a Wheat Crop – for Poultry – and for Swine. I am so totally ignorant of the nature of the Terms etc. to Mr Graham, that I shall request you your instruction, Mr Richmond is with me and I have and shall accommodate him with a Horse, He lives with my family etc.

I have received a Letter from Mr Graham (which I shall answer in a few days) relative to some observations and queries which he has requested my answer to. I think Mr Richmond's time is streigned, it ought to have been longer, but I am satisfied He will make great progress, for he have very clear conceptions of the subject, and draws very proper conclusions. I have desired him to answer Mr Graham's Questions by way of trying his judgment and shall explain any Errors he may commit etc. etc.

He was yesterday the 16th all the morning in the meadows, in preference to going with me to Dorchester Market, which I found dropping. Wheat rather above 8s. Barley 26s and 27s per Quarter

Nine Gallons – I sold last Week and this about 70 Quarters at 26s 6d and 27s. Oats are 17s to 20s per Quarter, our Crops are not good – Wheat not above 15 Bushel per Acre computed measure. Barley about 20 or hardly as much – Beef 3½d. Mutton 4d to 4½d Veal 5d. Pork 5s to 5s 6d per Score. Mr Richmond has made converts of two of our principal Farmers to the drill Husbandry with respect to the manuring in the Drills only for Turnips but all those things must be left till you see him. Coming at Christmas, I have had an opportunity of introducing him into many families, and next Wednesday we are to dine with, I think, one of the very first Farmers (though young) both for abilities or fortune, and who has I think the prettiest Ewe in the Country, and [his] Farmer, with whom we dined last Week (a man of very large fortune) who has the handsomest Dairy Cows in the West, what are called the true Devonshires – Mr Bakewell saw them when here, but since that they are very much improved.

I have not yet had an opportunity of shewing the Method of setting a New Ware, the Waters being in general so high that it is probable none will be set this Season, but I have shewn him one which I had set new this fall of the year – and I will have one of mine removed, if no other means can be found. Our Meadows look delightfully, having had no frost this Season – I think I can say he has, from the Hills, seen a thousand Acres under the watering system, and have already been in some hundred Acres himself. He is here at the wrong Season to see our Dairys – but just in the time for the Lambing Season, which Mr Graham wishes him to attend to. It will be useful for him to know – though only to lament, the difference of the Climate when in Scotland. I believe it's not possible to get your Lambs to fall so soon – many of our Farms are all in.

By Mr Richmond's account we have a much better method of cutting our young and old *Rams* than you have – We scarce ever [lose] one by that operation. He will take minutes of it. I cannot [see] any reason why you should not fold Your Stock Sheep more th[an] you do. We fold every age constantly except the Lambs and fatting Sheep and I see no good reason why we don't [fold] them. In Dorsetshire We don't fold our Ewes when growing big with Lamb. In Wiltshire they Lamb in the fold, and are healthy and well. Mr Bailey recommend me to get a Winnowing

146

machine, I will thank you to *tell* me the price and whether it takes up much room. We can winnow 80 Bushels without a machine for 3s and I apprehend you have two or three people employed with a machine.

January 21st Yesterday We were over two very Capital Farms, not so large as yours – the Stock Mr R[ichmond] will tell you his opinion of etc. when He see you again. Mr Graham I shall write to in a few days, but shall either leave it to his own Generosity or refer him to you.

Letter 10   11, 12 and 13 December 1790

Your concern for my silence is really flattering in the extreme to me, though you were quite mistaken in the reason that you ascribe for it. You don't know me indeed; to impute a mercenary cause to my motive. Had I never heard from Mr Graham, it would not have lessened my esteem for Mr Cully, nor for Mr Richmond whom I sincerely esteem. A Cause much more serious and distressing to me prevented me from writing or almost doing anything for a long time with any inclination or spirit. My eldest son whom Harry well knew was seized in June with a disorder that in July fell upon his brain and so deranged him, that He was obliged to be sent to a place of confinement – before it broke out so much as to be clearly distinguished, He had made away with a considerable matter, that with the heavy charges where he was put, rather embereassed me. He is now pronounced quite well and has been sometime, but not having fixed on a part of the kingdom where to let him reside for a little time to strengthen his resolutions, I have not yet removed him, though he is only a boarder there now. I need not say more for an excuse to a Person of your feelings. I could not say less, and yet remove every imputation of neglect, though it pains me much to relate it. I have been but twice to Dorchester Market since July, though only 5 miles off. I suffered severely by my imprudent orders, rather than trust to my man, at the Farm where I sowed the late Oats, by ordering them to be cut before they were ripe, for now they are threshing them (not with a machine) there is nothing hardly in them. The late Oats will not do for us for before our Barley is in the

147

Gentlemen sportsmen who are the greatest tyrants in the Kingdom are over the fields every day and if any late Corn is out uncut they are beating for Game all over it. The Wheat in this Country is by no means good this last year. Your Country loss I lament, but it staggers 12 Bushels Wheat per Acre is a great deal – when compared with my Crop at Piddletown where after the Tithe was taken away – I had barely fifteen Bushels per Computed Acre left for my self – and yet the land would let for 20s for such Acre. I congratulate you on your success with your Rams (I think if I know my own heart, there is not a spark of Envy in it) – it is really incredible here. We have not a Farmer in the County I believe that would give Twenty Guineas, or hardly half the money for the best Ram in England. The old Ewes at Weyhill fair (the most famous West of London) of which there are supposed to be 50,000 sold not so well as last year by two shillings per head the best – middle ones 4s per head and the lower sorts 6s and 7s per head less viz., from 14s to 28s per head. The latter fairs were better; Wether Sheep were rather low – Fat beasts are dear 7s per Score. Mutton 4d per lb.

Wheat seems to be getting up from 6s 6d to 7s per Bushel, our measure more than 9 Gallons – Barley I sold some for 27s 6d and Saturday 50 Quarters at 26s or rather more per Quarter delivered to Weymouth 13 miles – Old Oats I bought a few for a Gentleman 30s per Quarter; but 'twas thought very imposing – New Oats are now about 18s per Quarter; our Oats are never very good – I grew about 1600 Bushels Potatoes last season. I am fattening 5 Beasts with one sort of them, the others will be for sale – though at present very little demand unless the Sheep should want them. I am satisfied, Potatoes are excellent food for cattle that chew the Cud – Turnips this Season are a good crop but not large I have not yet weighed any, but intend it, if I have time and resolution – the drilled as well as the broad cast – I have not housed any turnips yet, perhaps I shall not this season – I don't know whether Mr Richmond explained the manner of doing it – lest he did not I will.

Cut off the roots, and tops (Greens) very close to the Turnip, which may be given to the Stock Sheep or beasts. Then having fixed on a dry spot near where the turnips are to be eaten – hurdles or a slight hedge is made as long as is wanted on each side of

148

a space six feet wide into which the Turnips are thrown and piled up as high as you think proper and as long as will hold all the turnips you intend to put – then are well covered with straw, then another Hedge or paling is made on the out side of the other, both sides and ends, and as high or higher than the inner one, two feet off from the inner one. This space is filled with straw all round, well trodden in, and also over the Turnips, like the ridge of an house, the whole is then thatched over, and it is finished – you begin when you want to use them at one End and take them out as you want them – I am the more particular in mentioning it – as it is now full time, or rather should have been done before; because by the word *housing them*, you might be induced to put them in to a spare house – Turnips will not keep in a house – they must have open air, only protected from the frost – this the straw does – in Houses they will taint and rot. – To change the subject, after a series of dry weather, We have had and still have a good deal of rain, very advantageous to our water meadows, I have a good many Acres now under water – this day has been one contin-ued rain and now ten o'clock at night still very violent. How boun-tiful is Providence to us – that even the rains and storms are made subservient to our use, if we receive good from his hand, shall we not also receive what we call evils, which repining, be it so – I sub-mit. How do your good Friend at Lord Tankeville's do, pray make my best compliments – I have likely had an account of their *wild Cows*. Pray do you, or have you introduced Cabbages into your Husbandry. The Marquis of Bath's Steward (forgive me for nam-ing Great folks but I am not a little acquainted with him) has done great things with them, he has raised them above 60 pounds weight each. My soil is not deep, nor good enough – Turnips are my object, as well as my best Neighbour's, I should often refer you to names was Mr Richmond at your elbow – Pray how does [?he] do, I owe him a Letter both of civility and thanks, but I am now out of the way of *Franks*. I once had my day, but it's over. Mr Graham sent me one and so did Mr R – just at the dissolution of Parliament – Mr Graham cost 2s 4d – I did not acknowledge the receipt of it to Mr G – because I thought the manner of his expression was different from my friend Mr Culley's. But *great* folks are not *like* little ones.

When you write to Mr R – make my best respects to him, and tell him I shall be very glad to hear from him; and I will certainly answer his Letter – if he say I've not done that – the cause you know – besides he wrote with a Frank I now cannot. Should I, which is now less likely than ever – or any belonging to me, ever go North – you and He would both be certainly called upon – Business engross your time as well as other folks, or – 2 days Coach from Loughborough, put you down at Piddletown – I am sensible that two and two make 4 days expensive travelling, and what is still more it is 4 days time, as well as all the extra ones – should you ever come West – don't come to fetch fire – rather stay away, than leave me to reflect upon my loss. – Christmas draws near – many and happy may the Holy days be to you all – I must draw *a veil* this Year over mine.

Letter 11   2 October 1792

You and Your Friend Mr Baily deserve my warmest thanks for your remembering this torpid part of the Kingdom. Whilst you are all bustle in dividing Commons, parceling barren Island, for *Holy* things may for a time be unprofitable and cast before swine yet in the End the Pearl of much value will be found. We are scarcly able to keep ours in the state our forefathers left them to us. The influence of *Majestic rays* will scarcely kindle the dormant spark in the Farmer's breasts, for though in his going to see the Farms round Weymouth he observed and told them "bad Farmers, bad bad Farmers, thistles, thistles thistles Farmer, earth hills, earth hills, Farmer everywhere, they should be cut, yes cut cut and kept neat." (This pray tell Messrs. Jobson and Co. he said to *Mr Bridge* where he went to see a fine Bull and some Leicestershire Ewes and Rams) yet thistles and earth (ant) hills still grow at *Winford.* The spirit for second or third rate Ewes from Leicestershire seems to be creeping out of it's sluggish bed. Your monopoly have not my approbation, circumstances unknown may support it but I know of only one or two that can be justified. I am sensible the Amor Patriae is a pretty fine thing to talk of, but the Love of one's self if not carried too far is the proper thing to support our necessary expenses and provide for our families. Perhaps the best method is to unite or blend them together so as to answer both purposes at once. Our friend Bakewell is an instance in point – I wish He may not at last, by a too eager desire to reallize a fortune, lose that Character which He for so many years merited, The friend of his Country.

I will thank you to direct the Oats to me thus GB Piddle^n. to Cotton's Wharf London to be forwarded by a Poole Vessel. You will favor me with a Line and a Bill of Parcels of the amount. I am obliged to you for what You have told me about the Drill Turnips. I shall next drill the greatest part, but one of the Plagues of Egypt we have that you escape – the Black Fly, an insect so small as to be but barely seen, that destroys numbers of Acres just as they come up. I have suffered by them in my drills this year, though Dung was put under every drill. The Drill turnip hoeing this year cost me 3s 6d per Acre – but I must contrive a cheaper method next year suppose they are spaced out by 6 Inch Hoes and afterwards singled by hand by Women and Children. A double mould board plough made to contract or extend I find finish the last plowing well with one horse – pray do you sow the *Headlands*, (the part that the Horses turn out at the ends upon?) if you do, whether drilled or broadcast? at the same time or after the first time of ploughing. I find the Horse in turning out tread the headlands so much that the turnips are scarce worth anything and yet to leave them unsown seems a great waste of land – I have never tried to drill them. We have had such a series of wet Weather never experienced before ever since the beginning of Haymaking and it still continues. A great deal of Barley is still out and some wheat. I was fortunate enough to finish three weeks since. Our Crops will be very light, I have threshed some Wheat and find the quantity per Acre very little indeed. I intended to have sown *old Wheat*, but the new was so damp, that the demand for old is very great, the price 8s per Bushel or very near Winchester measure which after the 10th of this month we have agreed universally to use only. Barley, I have been selling Old at 28s per Quarter; but what I have left, expect 4s per Bushel for or nearly. Oats (old) are very dear, I can scarcely ever grow enough for my own Horses, from 24s to 26s per Quarter. Beef, mutton and Veal in market 4½d the latter 5d, Pork 4d to 4¼d by the Pig. Butter 10d per lb. 18oz. and Cheese very dear indeed. What shall we say to all these things, and where will it end? I have some fears that the lower class will not sit still easy. We are not populous enough here. The fanatic Spirit of [them] have been but little broached here. I have lately heard from Nottingham [that] they had about half done harvest, and from Newcastle not a quarter done. We are as much inclined to wet now as we were three months ago, yet it has never been so very heavy as to cause any high floods. We have not heard from Mr Richmond? The best time for a person to see our Water meadows who can only see them, en passant, is in March not later perhaps the end of February? I shall always be glad to see any friend of Yours, but why send a *Substitute*. I am of opinion that Wheat and Rape will succeed, this Year was not favorable – indeed I have seen the rock I split upon; the Land was too loose, it should have been well trodden after it was ploughed. – I have not yet fed it off, shall judge better next summer than now.

The North Country travellers are often enquired after. I believe I told you the D. of Newcastle's master of his works has been here. He is doing a great deal in the Watering way and in a Capital style. I have seen a plan of his works, they stood in need of very little alterations – The only thing, I desired him *always* to remember, *Secure as much water as possible*, you can always let it away.

Our Family's best wishes attend the two travellers. My best Compliments are always with them and theirs, and truly so with You and Yours.

Letter 12   25 July 1793

Pray tell me the price of the Oats. *[written upside down]*
Though you may be surprized at my long delay in answering Yours, yet I flatter myself you will believe me when I tell you no one day has past without my wishing to write, some untowards circumstances prevented me from writing a more complimentary one, and the delay of a friend in Devonshire hindered me from performing your request satisfactorily – But now Your Friend Mr Mure is gone into the West. I need not be anxious about that or anything else in these parts, his attention, enquiries, and remarks will furnish you with much more information than it could be in my power to give You, particularly when you have it viva voce from him, and unless you have very particular engagements I sincerely wish you would postpone your publication till you *see* him, and possibly it will be for the interest of the parties to postpone it, for

I remember well *Lackington* says "in a time of War, nothing is read but politics".

To your first Question, We answer, there are scarce any genuine; the nearest to the original breed, are not delicate, but are hardy.

2nd The real Dorsetshires are short legged, and rather short in the Carcass.

3rd Average weighs about 18lbs. per Quarter. This County mostly breed, very few fattened here, the adjoyning Counties fatten our *Wethers* two years old. What *We* fatten are 4 and 5 years old. The oldest Mutton when fat esteemed the best in quality and flavor. The Ewes are all sold big with Lamb at Weyhill, October, to be fattened in Surrey round London; the Lambs for house Lamb – if well fattened not more than 1lb. per Quarter difference in the weight. Number of Ribs 13.

4th If Old esteemed excellent.   5th Average weight of the Fleece 3lbs. ½, price in 1792 14d per lb. but little ever sold before for more than 10½d.

6th Color, White, *white face, and Legs.* Staple good, Quality middling, Clothing.   7th from 3 to 4 years old! Tallow about 10 lbs. average.   8th only one at a birth, we wish for no more twins than to supply the loss by accidents. etc. – from the beginning of October to the beginning of April – the older ages are the earliest, the young Ewes latest in general, though there are exceptions to each – very well covered when lambed.   9th Natural Grass for the Ewes in the summer, Winter Turnips and Hay. Spring, Water meadow grass where it can be had. Wethers natural Grass. Fetches, Turnips and Hay, constantly folded except when Fattening. Ewes folded in Summer and Autumn.

10th A fatal disorder called the Goggles, no cure ever known, of a late years not so raging, and seems to be wearing off – no rational account given of the Cause or origin.   11th No means ever employed – except by crossing the Ewes with Wiltshires or Hampshires; either of which in many instances worse than our own. – The breed have been increased in size by this crossing, Turnips have increased the quantity and length of the Wool, but lessened the Quality, by making it longer and coarser.

The Dartmoor breed of Sheep I can speak nothing of but from hearsay and as your friend will be there, His intelligence on the spot, with his own observations will infinitely exceed anything I can say on that subject. To him I must refer you for that truly valuable breed of Beasts – The Devonshire, especially those near South Moulton. On looking over my memorandams, as well as the information from my Friends here this is the best account I can give You. And I am sure you will say, "And this is not worth sending". I own it. And to convince you how ignorant we are, In a large Company of Farmers got together for this purpose, not one could tell how many Ribbs a Sheep had! And the question was referred to the Butcher. From your mode of asking the question *I* was satisfied that different breeds – had different Numbers – but they would not admit of it – when I saw Mr Mace he satisfied me. My self not being in the breeding line had only general knowledge, but from many instances that occurred that day, I found my general, equall to their particular – for they knew nothing of the matter any more than myself. I am now about to apologize for the delay, but upon reading it over again, I find there is no need of any, for as it contains no *real, useful* information to you, it does not signifye how soon or how long it was before it reached you; except that you might have been eagerly, and anxiously expecting great and grand information from *Dorsetshire* : remember Aesop's Fable – a mountain in *labor*, a ridiculous little mouse *brought forth.*

Pray Sir, What kind of Folks are left behind in the North? Are we to judge by the samples you've sent us? Upon my word and credit we make a very ridiculous appearance, excepting one or two of my acquaintance, a Mr Bridge and Mr [?T] Banger whom your friends know – they all stared and looked like stuck piggs, with their mouths open.

I have only to request, you will continue me on your list of Acquaintance and let me often hear from You, and let me see as many of your Countrymen as you can spare to travel this way. What I have seen charm me, Though they are not *Bakewell's* and crusty, They are *Ready enough* and friendly, which I much more value – Two of my Neighbors were at the Dishley shew, in the beginning of the Month. But I fancy they preferred your *Homepierpoint* acquaintance. How is it in this obscure corner I pick up information of what's going forwards in distant parts? I startled my two Friends with the name of *Stubbings* as soon as they come back – though they meant nobody should know a word of it or him.

Will you believe me, I have more sincere trouble and concern upon my spirits and hands than I ever hope to experience again. The various Bankruptcys and stopping of principal Houses have so affected our concerns that we hardly knew which way to turn, and even now we are afraid it's not over. My next Neighbor and most intimate Friend (well known to Mr. Jobson and Thompson) Mr Masterman has lately stopped for £15,000. I had not the least suspicion of it. I am something considerable in, but not so much as (if it had happened a few years since) [I] have been. He will not pay more than 10s in the pound and that we must b*[illegible]* for. The Seasons with us have been very much like yours – I suffer upon a poor Farm, where my Son lives, very much in the Lent Crops. Barley very bad, Oats very short, unlucky both your sorts were sown there. The poland we shall begin to cut soon. I sowed them there because that Farm is tithe Free. Mr Mure told me of the Wiltshire method of using Potatoes. I shall try it – pray why *Would it not do for Turnips?* It was lucky I had drilled Turnips before this year. I am afraid I shall make but a poor figure this [year] Mr Mure assisted me in some particulars about the Dril. Messrs. Jobson's and Thompson's *Plough* is an excellent instrument, unluckily Mr Mures very short stay prevented him from seeing it in use. Indeed I blame myself exceedingly since for not setting it work – I quite despair of ever having the Shear properly repaired – I think Mr Thompson said that your *Smiths* had a mould to form them upon – is it practicable to get one? are they dear! or can our fools make another share if they had a mould to work it upon? Excuse my impertinencies. Corn has no price with us – that is there is scar[c]e any bought or sold. Wheat 6s, Oats 3s to 3s 6d per Bushel. Wool no price at all yet. Beef 5d and 6d per lb. Mutton 4½d and 5d Lamb very good 4½d Veal 4d and 5d (Butter 10d. 18 oz). Hay harvest nearly finished. The Early Water Meadow Crops good, the late ones rather light – I've just had a Letter from Sir John Sinclair acquainting me with the establishment of a Board of Agriculture, and with desiring me to attend it in London as they wished to try an experiment of watering Hyde Park and Saint James Park. I have not yet answered it – He is quite ignorant of my situation in Life, it will not suit my inclinations nor pocket to go two hundred miles at my expense to gratify the idle curiosity of every person that chuse to ask it – I have had one or two of those excursions already – *pro bono publico*, won't always do.

I very much doubt of the utility of those things in the hands of Lords and Dukes. Plain Country Farmers are not at *home* when they are with such sort of Folks – My hand, heart and Table such as it is are allways at the commands of *my friends* and nothing give me greater pleasure than to exchange mutual knowledge; but to dance attendance upon great Folks, and to answer such Questions as they may deign to ask you and then with an ungracious Nod be told you are done with – will not suit the Stomach of Your sincere Friend, An Englishman.

Letter 13   14 July 1795

I am favoured with Your obliging Letter, and take shame to my self as part of Your rebuke is true, though not so long since as you state it, as I wrote in the beginning of March. But I will tell you truly and honestly that I was ashamed to write, because from some very untoward circumstances I was not able to make you a remittance, believe me I have, from the almost total failure of last year's Crops of Corn and Turnips been really distrested. I have not had a Bushel of Wheat to sell since last February (very many are in my situation) except to our laborers and those many of us at County meeting almost two years since agreed to supply them at 5s per Bushel, which with a produce of not 12 Bushels per Acre must be a severe loss you can easily conceive. In almost 40 Acres of Peas and fetches I had not 2 *Bushels* produce in the whole. Great part of the Sheep that should have been fattened are now in as bad state as they were 12 Months ago. I could say more, but it's useless and wrong to complain. The Hand that gives can take away; and let us hope, will restore again, to add to the Country's calamity, the 18th and 19th June such heavy (cold in the extreme) rain fell, that a very considerable number of Sheep were killed by it, being in the midst of the Sheep shearing, many Farmers in this neighborhood lost some 10, 20, 40, 60, 80, 100 and one near 200, out of 240, being killed by the Severity of the Weather. I had not *shorn* and therefore lost none.

To change the dreary scene and advert to a more pleasing one, Your success in Your Sheep markets astonish me. We are wrong,

156

that's certain: but how to amend is the Question, and as the poet says, there's the rub! I spake not of myself for I am not situated on a breeding Farm, our Farmers are great breeders; but can do nothing like it at all. Indeed our Sheep, I mean wethers, seldom to go off to be fattened till four Years old. In the rich land in Somersetshire they are fattened at two years and half old. The few I made something like fat 4 years old past, averaged about a Guinea and half, which we think good work. You say, at 60 You can't positively say You shall be able to reach one extra hundred miles, beyond the spot where You are to pay for an Estate gained I hope by your industry. How can I at 60 stir from home, without having scarce done *any* thing to that Age – but I don't repine, and at all adventure let me know what the peas come to which with the Oats I will certainly pay you for as soon as I can. We have been truly alarmed here, for I am certain, if we had not had two Vessels laden with Wheat come into Weymouth 3 Weeks since we should for more than a Week past been without one handful of Wheat or Barley in the County, even now our poor are for a day together in many Villages without a morcel once or twice a week, and we are obliged to support them – though in the trade I am in, Grocer and Mercer's Shop, there is scarce any money to be taken. We have had private and County subscriptions for them. Before, the poor (and the Rich too) would believe there was a scarcity they were troublesome and in the neighborhood riotous, but now they are decent and orderly, and are thankful to get it. I pay the extra price above 16d the half peck loaf to all my laborers Men and Women, which is now 4d and expected to be more. God knows where it will end, three Weeks more will give us new Corn, our Crops are promising. Turnips likely to be very good, Hoeing is begun. Hay making almost half over, light crops, partly owing to the very cold weather, and partly to feeding the water Meadows so late. I had about 30 hundred weight of old water meadow Hay left, which my Horses ought to have had, but necessity ha*[manuscript torn]* I sold it for 5s 6d per hundred weight. We have no markets *[manuscript torn]* Wheat 10s 6d to 11s Winchester. Barley 5s, Oats 28s per Quarter. Beef alive 9s to 10s per Score. Mutton 5½ per lb. sink offal. Pork 9s 6d per Score (20 lbs.). Butter 10d and 11d per lb. 18ozs. We have had a Land surveyor over a farm I rent, a worthy good Landlord died, left a very Young successor, whose Trustee's are trusting [?thrusting] their noses everywhere. I have 7 or 8 years to come so am not very anxious. The Young ones must do for themselves, if I can't do for them. A Mr Richardson was the valuer. You know something of him. He knows something of Mr Bailey – He is now a great man etc. etc. My best respects to all friends, pray let me hear from You soon, and assure Yourself I will not be ungrateful.

The pease though unfortunately came so late make at present a decent appearance, they are coming into blossom and I hope the season will be favorable to them, for as you describe them to be an early sort, they will be valuable to us here. You mentioned sowing wheat in the spring of the Year, pray is it a particular sort, or will the same kind that we sow in the Autumn, do to sow in the Spring. Your *Church's* Oats in good land keep their quality that is if the quality one year is not so good, the next if a favorable season, bring them to the first standard again – I have read with pleasure your treatise, that *unknown* something in the handling is still *a mistery.*

Letter 14   7 January 1796

I acknowledge the receipt of Your favor, with pleasure because it tells me the three travellers got home safe. But I can safely say, I never was so much disappointed as with Your departure, an acquaintance of so many Years and at 600 miles distant, to which the several obligations I lie under (that will never be forgotten) are to be added and all could not procure more than two short winter days of Your company, as often as I think of it, I almost wish I had not seen You at all, for then I should still live in hopes, but now must despair of *ever* having that favor again. A serious reflection arises here, but I will suppress it, and go on with telling you Mr Banger came in half an hour after you were gone to conduct you himself to Druce [Farm].

Mr Bridge (whom Mr Jobson knows) was so hurt at hearing you had been in the Country, and not to see you. He was inclined to have been angry with me (for he said if he had known it, He would have come at midnight) till I told him how short you[r] stay was, and that I could not dream of your going so soon, nor

158

could I say to a Friend, as soon as I saw him "when do you go away". They are so compleatly disgusted with the *manner* of Bakewell and Honeybourne, that, they wished to have experienced the open, ingenuous manners of the North. I had a thousand questions to have asked you, as well as a candid acknowledged [?acknowledgement] of the obligation I feel relative to the Peas, they will not be forgotten and when the debt's discharged the obligation will not be cancelled.

I am fond of but few acquaintance here, therefore can the easier select them: all, or most of them have read your Treatise on Stock; and are highly pleased, the two I mentioned could have given you much information relative to the Dorsetshire stock. I have some Devonshires that I should not have been much ashamed to have let you have seen, but the moment you announced your determined resolution, you quite discomposed me, I say it, that if you say any thing in me improper afterwards, or not so hearty as you might have expected, do let me intreat you to place it to that account, for it is the real truth. I shall take the liberty of writing to Mr Bailey, if I shall not be thought impertinent upon the subject of his Drill and plough. He obliged me to look over my school books again, e'er I could fully enter into the subject, by the help of them *Rowney* and *Nicholson* I think clearly understand his treatise. At the Bath society (which met the tuesday after you left us), A Threshing machine was exhibited to work by two men by means of Winches. One of my Friends is to have one, and I shall then see it, and shall afterwards trouble Mr Bailey upon the subject. It will thresh about 50 Bushels of Wheat in a day, and is portable, the only good quality belonging to it. Should you see him before I write pray ask him "whether a Cog wheel of from two to three feet diameter (turned by hand) working a pinion on whose axis is a cog wheel that work in a pinion on whose axis is the Wheel, to whose circumferance the switchers are fixed can be turned by *one person* with a sufficient velocity, for instance, 2000 feet in a minute? If it can I have not a moment's doubt of making one from the drawing he sent me some years since, and which I keep by me as a valuable prize. I suppose a man can turn a *winch 40* rounds in a minute, if so, we have a data to go by. This person has no patent, pray does Mr Bailey mean to have a patent for his drill plough? If so, I shan't ask another question about it, I think he said to me, he did not intend it. It is the simplest and best I have ever met with, and with a little of his assistance can easily have one made I believe. Wheat is worth 10s 6d to 11s 6d per Bushel Winchester. Barley 33[s] to 35s per Quarter – Oats 26s to 28s – Beef 5d per lb. Mutton 5d and 5½d. Veal 6d. Pork 5d and 5½d. My sheep have had *no Hay* yet, and the T[urni]ps all carted off the land, but the shepherd think I am mad and that the Sheep will never get fat, yet I think they get forward apa[ce].

Had you been under 30 I should easily have conceived you meant to flatter Sarah, I thank you for your good opinion of her. She is nearly even with You, for had you been an unmarried Man, I don't know whether She would not have crept into your Saddlebags, she desires me to make her best respects and wishes and only lament she had not more opportunity of shewing her attention to her Father's friend.

It is with gratitude I say it, I have great reason to be thankful to Providence for many Blessings and Comforts, and *those* around me are not the least. Wishing You and Yours the Compliments of the season, and requesting my best respects to all I know.

We have had neither frost, snow nor hail yet, but abundance of Wind and stormy weather. The water meadows look as green as they did last year in April. The Lambs are late this Year – the forward Farms have had a great many in, and many twins. The Ewes and Lambs look remarkably healthy and well. The Wheat look very fine upon Grou[nd]. I have resolved to sow, or plant, or both some in February or March as the weather suits. The Bath Society have offered a premium for 5 or more Acres dibbled. Pray say to Messrs. Jobson and Thompson with my best Compliments, our Christmas is very dull, they would enliven it.

Letter 15   8 September 1796

Very often have the recollection of my being a Letter and many et ceteras in your debt determined me to write tomorrow and tomorrow which like the Fool's holyday still pass away and leave me as irrisolute as before, business might justly be pleaded as an excuse if the obligations of gratitude for favors received did not severely reproach with "no business is equal to friendship". I told Mr Bailey I would write to you and my daughter has more than once asked me when I did write to Mr C — to make her best Compliments and say, "She will never forget You". I have lost some time since the Friend of my bosom, and the soother of my Cares my Children's *Mother,* but my *wife's* sister whose ill state of Health and gradual decline prepared me for the stroke, yet when it came I felt it very seriously – for new friendships after Sixty are not very desirable or easily formed. Since that a Brother in law suddenly taken away in Norfolk, and very lately the oldest relative I had and the nearest, (my Father's sister) at almost 90 closes the present bill of mortality and leaves the memento, that We are soon to follow. Mr Bailey's machine for drilling Turnips I have used this Season and the first part of it with great success. The middle and latter sowing failed a good deal, and one thing I am much at a loss to account for every alternate dril was much thicker than the other, the mode of sowing was this; I was apprehensive the roller going twice over the ground would press it too hard, therefore every other drill was made with the drill annexed to the roller, and then the other alternate ones made by the drill (drawn by the horse) without the roller, now having no Idea of any such thing happening, I cannot speak with certainty which are thickest, but there is a material difference. Did you ever observe the same thing? Or which way do you dril yours? The Instrument is infinitely the best of all that I have seen, and the principal just.

The Harvest (Sunday the 11th) will be nearly all in this Week, and the Crops abundant. My Spring Wheat (sown too late I believe) is getting ripe apace, but the straw of *that only* is very much *struck* and I beleive the Corn blighted – and if so, it is really the *only* instance. The prices of Corn we know nothing of scarcely now for every one is employed in (and some have just finished) Harvest, but assuredly every species of Grain must sink greatly. The Red Clovers are vastly well seeded and round us never so much saved for seed. Should the autumn be favorable it must be cheap. There will be none, with us saved next year, for there was hardly any sown, from it's price and scarcity this Spring. Wheat is worth 7s per Bushel. Oats 3s. Barley no price yet. Beef retails at 6d per lb. Mutton the same. Veal (never finer nor more) at 6d. Lamb 5d. Pork 8s per Score. Our Hay crops were plentiful, and at the latter season a fine time for making it. The Harvest season was never known finer and is now quite summer weather.

Mr A. Young has been in these parts, called here, but I was out – called upon Mr Davis at Long Leat, the Marquis of Bath Steward. I wish Mr Bailey had seen Mr Davis, he would have found him truly a Man of Business, the Marquis's large concerns all pass through his hands. Mr Davis was not at home when Mr A. Y[oung] called on him, nor did he regret it!

The Gentlemen of this County have fixed on me, to correct the Survey of Dorset. At present there is some demur on my side. I suppose You have seen the circular Letter sent by the Board, or their representative, for We can scarce believe the real Members of the Board would have written such one, saying the Board would be at no farther expense. I told our Gentlemen I wished to see my way clear before I went any farther, that is, a promise of my real expenses being repaid me; here it rest. In the mean time they have desired Mr Young to send me the necessary papers of information from different people which the Board had received; not one of which he thought proper to send, witheld them and sent sketches of heads of Chapters to be filled up by the reporter: that might enable *some other person* to draw up a private account previous to the correcter's report. Mr Davis was served in this manner in his Wiltshire report, without any acknowledgement from whence the information came. I smelt a rat and stand aloof. Pray how do all my acquaintance in the North. I shall not repeat to You what I said to Mr Bailey relative to Young Church. He is a very extraordinary Young Man. I think it was a pity he had not had a few Years more experience before he made the tour; had his judgment been more matured he perhaps might have profitted more, but he is certainly a wonderful young Man, and 'take him all and all, You will not quickly find his like again'. My respectful Compliments to all, Mr Jobson and Mr Thompson at the head of the List. Pray make my best respects to Mr Bailey and say, I much *[manuscript torn]* to hear from him; I dare not make that request to You, *[manuscript torn]* only say, the oftener I hear from You, the more you incre[ase] the obligations I feel myself under, and nothing, but a firm resolve of acquitting them, enable me to say a word on that subject. Politics now engross the general conversation and I much doubt whether more than words won't be necessary: it is rather late in life to take the musket into hand, for like old Priam's it would be a feeble effort. One Son they have in the

Volunteer cause, and the other should not be withheld in the time of real danger. For I think I could say with the old Patriarch 'if I am bereaved etc.'. There is a female relative now at my elbow whom you know, that would cause many a secret twinge to leave her unprotected, but hold *enough of that*. There is a secret providence that watches over and protects and guards defenceless Innocence and there we will trust her. My best wishes attend all.

Being too late for the post, I will just add that our Encampment between this place and Weymouth w[h]ere the King has been some time fill up the void space with numbers who do not know how to spend their time. Perhaps you will say we are of the Number. This day (Tuesday) myself and my daughter went to see a field day as it's called – We went away before Ten and home by four, all which time She was on Horseback, and brought a good stomach home with her then – no refreshment between. There were a large concourse of people. It gave me pleasure to see even the appearance of Loyalty in every face. They leave us the end of the Week, and then the *farce* will end for the present.

I am amongst the number of those who really rejoyce to see the King look so well and the Whole Family of Females so happy. May his day be many and his years in *Peace*.

Letter 16   6 July 1804

> No business done at the Fair. 7th and 9th a torrent of
> Rain. Hay spoiled seed Ray and trefoil in a very bad state.
> *[written upside down]*

An increase in You will I know be an increase of happiness and pleasure to You, because it will, if possible, add to Your amiable daughter's, whom I now congratulate on the secret inward Satisfaction she must feel on being a Mother – she who was ever amiable as a daughter to you, will now have a double claim on You in the light You now see her in. Make our best congratulations to her and all Your family on the joyful event, my wish and prayer are that they may be both a Comfort to you in your declining Years.

Long have we been awake to Wars Alarms! When shall we sleep in the bosom of Peace? is a serious and important Question. Till then all that can be Soldiers must be so or the event will be fatal. I with you, should have thought Mr William Jobson almost the last man, that would have weilded the Sabre, next him I should have scarce believed my son Samuel would ever have put on a Red Coat, yet he has with the Company been on permanent duty at Weymouth for a month, without being one day absent, or once home. I had six of my own men doing duty in it all the time; the day they marched to Weymouth was the day of the alarm of the French being landed in Portland; they were ordered to march as expeditiously as possible with twenty rounds of Ball Cartridges per man; they gave three Huzzas and marched off from Dorchester within half an hour. It turned out a false alarm. The Season with us has been very untoward. A continued drought for six weeks and the last fortnight very hot and parching, has burnt up the Grass – the late sown Oats and Barley, and also all the Corn that was sown on light thin sandy lands was going off very fast indeed. – The watered meadows having had abundance of water in the early part of the season, grew very fast in the hot weather, and is a great burden, and so are the Clovers – we began mowing Clovers the 17th ultimo and by last Monday night we had gotten the Clovers secured, and some water meadow hay. Since that night we have had very fine rain and some part of the time very heavy one meadow near Park house which H. Rutherford knows, is 9 Acres, and George Hoar and another were 5 days cutting it. It was yesterday the greatest part underwater, if we should have good weather now, little will be spoiled and not less than two Wagon loads an Acre. No Turnips stand yet, our People hereabouts began sowing very early this season but they must sow again. I have drilled some the week before the last, and have met the same fate, so that what should have been first will be last. Another fortnight will determine our fate, both as to Corn and turnips – the season is such that I think we must either spoil our hay, or lose our Corn and turnips – both I think cannot be saved.

This day is our first Lamb fair and also shorn wethers, I shall hear by and by the event. But I think they will be dull I had but a small lot of wethers this year 50. They sold *altogether* out of their Wool for 59s per head – but they were very good and fat, under 22 lb. per Quarter. Wheat sinks, from 6s to 6s 6d per Bushel if good – Barley 3s 6d to 3s 9d but little left in the Country. Oats 3s to 3s 6d per Bushel: Beef scarce and dear 12s to 14s per Score. Mutton in market 8d per lb. Lamb 8d. Veal 8d.

Mr Banger is just returned from Woburn. As he was a stranger there, He begged me to give him the address of my North friends, if any of them should be there, but I find by him You were none of you there – Sir John was not there neither. He ea *[manuscript torn]* got introduced by some friends which He *[manuscript torn]* by that means found it very agreeable, the Duke was very attentive to every person; yet it plainly appeared it was not the *late Duke*.

6th in the Evening. Fine and clear. Should it now continue the value of the late rains will be incalculable and the damage to the Hay but trifling. We seem to be full of sheep and Lambs stock, and had not the Navy at Portsmouth and Plymouth taking away

[A page of the original letter has obviously been lost.
The next surviving page continues as follows.]

let me request You to make our best respects to all and every part of Your good family. Sally is by no means well, and yet tolerably full of Spirits, too much indeed for her strength, her complaint has lately been rheumatic particularly in her loins, so that she can take but little exercize by walking, and none by riding. She has been at a friends for a week and is just returned.

I shall wish for 5 Quarters Pol[and] Oats early

Letter 17   10 December 1805

Please to direct – Piddletown Dorchester Dorset our post office is changed *[written upside down]*

To attone for my past negligences I take up my pen, to enquire after Your health, together with Mrs Culley, Your good sisters and son as well as those of your *Darlings,* and that the new Year may bring with it all the blessings a good and gracious Providence has in store for You. Just after I wrote last my daughter had either from a violent cold or a return of her former disorder, a most severe pullback, which reduced her to the same state as she was in when she went to Charmouth, and though she staid there for three weeks more was not able to Bathe again till three or four days before her return. Since which time she has experienced the visitudes of one day better and another worse, till within the last two three days she has again been better, and yesterday walked almost two miles, promising herself such another today, but early this morning the Elements were at variance with her intentions and a heavy torrent of rain gained a complete victory over her, and enabled me to write two or three Letters – to say, that though she was disappointed I was not, after it did come, for making a virtue of a necessity I rejoyced to think we were going to have a plenty of water for our meadows. Never did I know a finer Autumn for eating the aftermath. Our Cattle are scarcely yet come into the straw barton, or the Milk Cows to Hay. As I shall probably not send off this for two or three days I will leave room to say the event of the rain's continuance or not. We are, speaking in general, disappointed in the produce of our Wheat crops as well as in the Quality which are both but indifferent, though there is a good deal very good and tolerably abundant. The price is fell (and still sinking) from 10s per Bushel to 9s the very best and inferior will not bring more than 8s – Barley crops were good, but as the only rain we had during the latter part of the summer fell in the Barley harvest, a good deal come to hand cold and damp. The increase to the maltster will be small and much care required to bring the malt good to market. Price was 46s but now not more than 38s and still sinking and I believe will very soon be down to 32s per Quarter, Winchester. The Wheat seed time was very fine and has been some time over except with a few individuals, who are sowing quantity after Turnips. Can you tell me why I cannot or rather how I shall sow wheat after turnips in the spring Season with success? I have tried more than once, but not with success. I much wish it for as a Maltster I had rather sow wheat than Barley. In other words I had rather buy my Barley for making Malt than grow it myself.

I don't know whether I have not lately made a bad bargain, but I did it with my Eyes open. I have taken about one hundred acres of enclosed poorland, and yet there are a number (being in small fields) of as good thorn hedges growing upon it, many 10 feet high as I have seen any where – the land is very poor and has be[en] beggared out. I am to build a House and offices and at m[y] own expense, rent free tithe included except 20s per Annum in acknowledgement for Twenty one Years to come. Chalk I have near to use in it's crude state or to burn into lime –

the latter I prefer on account of the less Quantity of loads necessary, my principal motive was as an amusement for my Son and
for his health's sake being about four miles off. I am intersecting
it with open drains, it lie very well to drain but its subsoil is a
strong hungry clay. Last Thursday was *truly* a day of thanksgiving
to us all. If Peace succeed it will be happy for us all.

Tuesday morning 17th. The weather changed again from wet
to dry with a sharp frost, which has continued increase ever since
and we are now quite locked up. I feel the cold more than ever
and can scarce pluck up resolution to walk out in it. We all go on
just the same, one day better the next not so well. I led you into
an error relative to Samuel for though a certain Young Lady was
with his sister and came from thence to Piddletown where she
now is with me, yet I believe till peace returns, she must remain as
she is.

God bless you all Prays faithfully, Your sincere George Boswell.

NOTES ON THE LETTERS

Letter 1: The 'friend' referred to here is presumably Robert
Bakewell, who had recommended that George Culley should
write to George Boswell to enquire about the method of watering
meadows.

Letter 4: The reference in the first paragraph is to George
Boswell's wife, Sarah, who died in 1777 and to whom he was married for seventeen years. 'Harry' refers to Harry Rutherford, who
had been sent to Dorset by George Culley to learn the method of
watering meadows.

Letter 6: The bishop of Llandaff from 1782 to 1816 was the
Right Reverend Richard Watson. He seldom visited his diocese
but lived at Calgarth Park on the banks of Lake Windermere,
where he devoted himself to experimental agriculture and the
improvement of his estate. Earlier, he had been appointed
Professor of Chemistry at Cambridge although he confessed that
it was a study of which he 'knew nothing at all, had never read a
syllable on the subject nor seen a single experiment in it'. He did,
however, contrive to lecture on it, and later secured the Chair of
Divinity at Cambridge. The 'clergyman from Gloucestershire' was

the Reverend T. Wright, who had written *An Account of the Advantages and Method of Watering Meadows by Art* (Cirencester, 1789, Second Edition 1790).

Letter 7: A second and enlarged edition of George Boswell's book on watering meadows was produced in 1790.

Letter 8: 'Mr Bailey' was John Bailey, estate agent to Lord Tankerville of Chillingham Castle, Northumberland. He joined with George Culley in writing Reports on the Agriculture of Northumberland, Cumberland and Westmorland for the Board of Agriculture in 1805.

Letter 10: The steward to the Marquess of Bath was Thomas Davis, a notable exponent of improved farming methods, whose report on the *Agriculture of Wiltshire* was published in 1794.

Letter 11: The reference in the second sentence is to the enclosure and cultivation of Holy Island. 'Majestic rays' refers to King George III's visits to farms in the vicinity of Weymouth, where he was staying on one of his regular visits to the town.

Letter 12: Sir John Sinclair was Chairman of the Board of Agriculture, which had been established in 1793.

Letter 14: The Bath Society refers to the Bath and West of England Agricultural Society, which had been founded in 1777 and was having a great influence upon west-country farming. Sarah was George Boswell's daughter, who kept house for him after the death of his sister-in-law in 1796.

Letter 15: 'Mr A. Young' refers to Arthur Young, one of the leading agricultural writers of the time and an influential publicist for improvements. The 'Survey of Dorset' refers to the *General view of Agriculture of the County of Dorset*, which was produced for the Board of Agriculture by the Dorset surveyor, John Claridge, in 1793. This attracted a good deal of criticism from local farmers and landowners and was eventually revised and enlarged by William Stevenson in 1812.

APPENDIX 1

THE WARNE FAMILY TREE

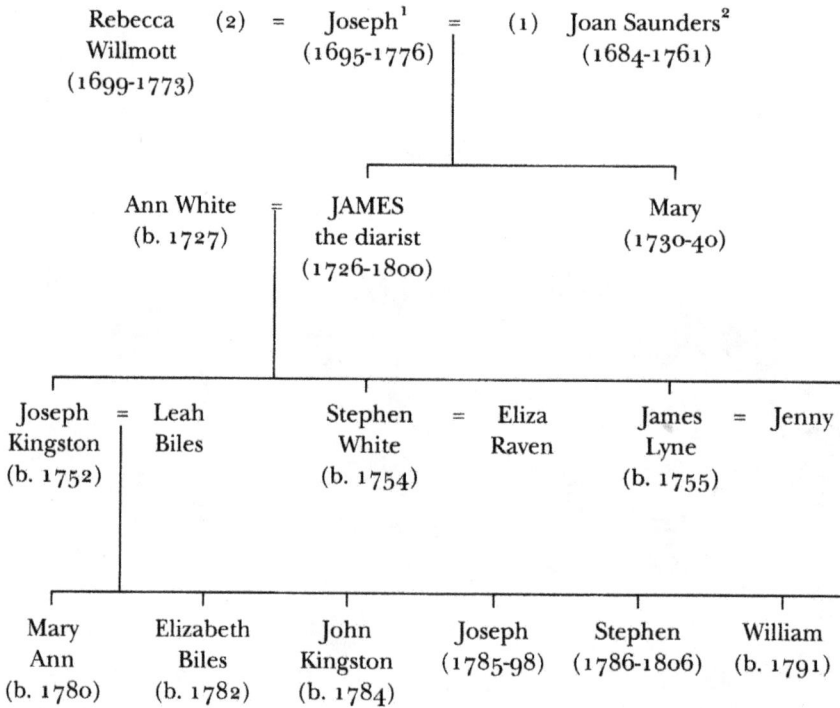

```
Rebecca      (2)   =   Joseph¹   =   (1)   Joan Saunders²
Willmott               (1695-1776)            (1684-1761)
(1699-1773)
                                   │
                     ┌─────────────┴─────────────┐
         Ann White = JAMES                       Mary
         (b. 1727)  the diarist                  (1730-40)
                    (1726-1800)
                    │
   ┌────────────────┼──────────────────────────────┐
Joseph   = Leah        Stephen  = Eliza      James   = Jenny
Kingston   Biles       White      Raven      Lyne
(b. 1752)              (b. 1754)             (b. 1755)
   │
┌──────┬─────────┬────────┬────────┬─────────┬──────────┐
Mary   Elizabeth John     Joseph   Stephen    William
Ann    Biles     Kingston (1785-98)(1786-1806)(b. 1791)
(b. 1780)(b. 1782)(b. 1784)
```

NOTES

1 Joseph, the diarist's father, had an elder brother, Christopher (1694–1755).
2 Joan, née Duell, was formerly married to William Saunders (1688–1722). The children of this marriage, the diarist's stepbrothers and stepsister, were William, Michael (married to Mary) and Elizabeth (married to Henry Loxly).
3 Charles, antiquary and archaeologist, was a friend and correspondent of William Barnes.

Robert Duell (b. 1757)	Anne (b. 1759)	Mary (b. 1761)	Christopher (b. 1763)	Elizabeth (b. 1764)	Martha (b.1766)	Roger (b. 1768)

Martha (b. 1792)	Charles (1794-97)	Robert (b. 1796)	Joseph (b. 1798)	Charles[3] (1801-87)

APPENDIX 2

APRIL, 4th Month, has xxx Days.				[Week 16]		
Account of Monies	Receiv'd			Paid		
	l.	s.	d.	l.	s.	d.

17 | Monday — Ho: pl'd w'th 3 Suls & Sow'd 12 B'l Barly in Hog lease, I & Nanny rode to Wa: Fair We saw Fr & Fs: While there We was at Br Florences & Din'd & Drank Tea, I Bo't a Book, The Art of English Poetry w'ch cost 12v3d y'e other book 12: & Large Bill Cattle sold

18 | Tuesday — of 7 Ho: Pl'd w'th 3 Suls & Sow'd 12 B'l more of Bar=ly. I rode to Pid & measur'd Hedges w'ch Charles Hodges have made & p'd him for it Nanny rode to Heffelton to see her mo'r & met Sister Florence there & as I came back from Pid: I call'd there, & We came home together ab't 8 at Night a dry F. N.

19 | Wednesday — G: Sinnick, W'm Locass & Robin came home w'th y'e Wagon &, 5 Ho: (y'e other 2 rested) &, carr'd a 10 Qr's of Raygrass Seed to Fw: Brimes at Chettle down, near Blandford they arrived 2 B 3. & set out ab't 4 W. Locass went &

20 | Thursday — W Locass set out early to Pid: w'th B'r M'c: Old Ho: & pl'd & Sow'd 4 B'l of Barly w'th 2 & G. Sinnick to R. Locass came home w'th y'e Wagon & 5 Ho: & carr'd 5 Sacks of Barly 2 or 3 of Oats & 10 of Raygrass Seed & F'rs: Old Roll: of Bov: w'ch was up there & when y'e Wagon departed I rode to Pid: & saw how

21 | Friday — 4 Ho: drag'd down Resters in Mountfield &3 made an End Plowing 2 Ridges y't was begun & Sow'd 4 B'l Barly & 36 of Grass seed & X Harrow'd it in w'ch is 2 y'e Lower Hoglease. In Allen put y'e new Bridges by Gully, I call'd at Heff: as I went to Pid: & had 36 of Clover Seed at 4d: & I call'd at Bov: as I came

22 | Saturday — 4 Ho: Drag'd as yesterday, & 4 Pl'd w'th 2 Suls & Sow'd 8 B'l Barly, I rode to Wa M'r w'th y'e Butter I ventur'd my Chance for a Packet of D'r Arford & a Silver punch Ladle he gave I't was got by a Man of Wa:

23 | Sunday — I & Nanny went to Wool Ch. in a N. Text Ps. 41. 4 & We Sung Ps. 128. 96. 117. M'r Fr: & Son all was as it I S. in his new Clothes, Rob't Standley's Wife & Rob't Stickland's Wife of Burton came w'th Nanny from Ch. & R. Standley came w'th me & R. Stick= land came & they all stay'd till ab't 8 Clock

Appointments in April, 1758.	MEMORANDUMS, or, OBSERVATIONS.

dear a Cloudy Morn: Rain all ye Middle of ye Day
& a fine Dry Even: to come home in Wt got home
abt. 8. Saw Mr. Harvey of Chichester at Wa: Fair.

but abt Nn. ye rain came on & raind till night

gardend for himself at Wool. I went to Bov: after
Dinner & got an Nt. 2 Ho. of Br. Mr. Saunders's & had
it shoed, & intend to drive him a little this season.
G. Northover. Jas. Bascomb & his son Jas. & my Fr. wth
them dug up abt. of 18 inches deep between to of ye
N. Pillars of ye Ch. a large Tomb Stone long — 7-5
things was there, I got home   Wide — 4
abt. 8. Far. Jarves pd. me 9-2-0   Thick — 0-6
ye award money towards mak My Fr. intend to beg it of Mr.
=king up ye Hedges there Mr. Wold.
Robinson his Wife & Br. Thos. & Far. Charles Burden & his Wife
came to see us abt. 5 oClock & staid till midnight,
a Blessed fine day Mild & Warm.

back & Saw my Fr. & Mor. Fr. intend to go to Blandford Mt
tomorrow to meet Uncle Ja: Snell there, Dr. James
came to Bov: just before 7. & I came away soon
after. mild weather some few drops of rain.
The Masons have got ye Old Kitchen at Mid: so low
down yt J. Allen. helpd them down wth ye Chimny piece yesterday.

I got home 8 Bg. fine mild pleasant weather
Wheat Sold for 6s. Buy & some 2d more.
I sold 10 Pints Kidney Bean's for 3d. a pint to Ja
Brown & left wth him fa Ho to sell for me 5 pints
Bunch Peas & 2 pints of Black Ey'd at 2d. a pint:
a very brisk drying wind at East. wd. get very
hot had there been no wind. I rec. a Lr.
from Mr. Kingston for my Fr. to meet him a
Lulworth Castle next Thursday morning.

GLOSSARY

barm	*yeast*
bay	*a river dam*
chilver hog	*a ewe lamb, unshorn or under two years old*
cockled the vetches	*weeding out the corncockle*
cross harrowing	*harrowing alternately at right angles*
dragged	*used a drag, a heavy harrow*
drat	*shaft of cart or droat, leather harness, also team of horses*
druggish	*slow or dull*
durns	*door jambs or gate posts*
dust	*meal dust or bran*
fan tackle	*winnowing fan, canvas sails mounted on a hand-driven rotating frame*
frazzle stick	*(probably) a frayed stick or one with twigs at end*
frith	*brushwood or underwood suitable for 'wreathing' or wattling*
fus	*gorse*
gads	*nails for spars, stake or bar of metal*
gave off	*finished work, knocked off*
hassocks	*tufts of reed or sedge*
Holliry	*Hilary day, 13 January*
kill calf	*a serious cattle disease probably caused by parasitic infection*
lambstones	*lamb's testicles*
lave	*bale, lade or draw water*
lug	*measure of land; in Dorset 15 feet and an inch according to Marshall in 1817*

174

main, new main	*water channels for meadows*
mall	*possibly marl, context is wrong for malt*
minice	*minnows*
mixon	*dunghill, heap of ashes, refuse*
mores	*roots or tap-roots*
mull	*dust or rubbish*
necessary house	*earth closet*
old Michaelmas day	*10 October, eleven days later than 29 September, owing to change of calendar in 1752*
posstick	*postic, i.e. rear shaft*
pooked	*made a pook or stook of corn*
prize	*to value, or to weigh*
quine	*corner*
raring	*usually to raise or finish (of building); here, probably, to add a new layer*
restering	*the action of using a wrest, i.e. tines fitted to plough to harrow more deeply*
rolling bay	*[?], see bay above*
rout	*grub up*
scroff	*piles of turves, or dead wood from under trees*
se'nnight (senit)	*one week, seven nights*
skilling	*outhouse, or lean-to*
spear	*reed stems or stalks*
spurred, spurring	*spread, as with manure*
stavel	*stone pillar*
sul(l)	*a plough*
tut job	*piece work*
wait	*weigh, a standard measurement for wool comprising 6½ tods of 1 stone each*
ware or wear	*weir*
yeograss	*aftermath or after-grass*

INDEX OF PERSONAL NAMES

Brown, Henry, 34, 43, 47, 52; James, 41; John, 22, 47; Mr. (*brother-in-law*), 13, 39, 59, 82, 88, 105; Mrs. (*sister-in-law*), 38, 39, 59, 76, 82; Mrs., 37; Sarah, 46, 50, 75; Ursula, 24
Brownjohn, Mr. (*gamekeeper*), 51, 108
Bungy, T., 110
Burden, Charles, 35, 41, 50; Philip, 24, 25, 26, 31, 33, 36, 81, 82; William, 53
Burgess, Jay, 104, 106
Burt, John, 56
Bushell, Richard, 110
Bysshe, Edward, 40

Cartwright, Francis, 15, 16, 32, 34, 35, 39, 47
Charity (*Fisher's maid*), 65
Chaffey, J., 94
Chapman, Sarah, 117, 118
Chip, Mr., 27
Chisman, Mr., 26, 30
Clark, Mr., 91
Clavell, George, 7
Clavells, Jenny, 44; Molly, 44
Clench, Mr., 27
Coaks, Luke, 111, 113; Shepherd, 110, 111
Cob (Cobb), Daniel, 23, 24, 26, 28 – 31, 34, 36, 37, 39, 44, 47, 49, 50, 53, 61, 64, 65, 70, 96; Jane, 49
Cockram, Mr., 29, 36, 48, 52, 65, 91, 94, 110, 112; Mrs., 29, 48
Coffin, Elizabeth, 48; Thomas, 21, 27
Cole, Admiral, 12, 26
Colhoun, Mr., 127, 132
Cooper, Mr., 87
Crocker, Robert, 61
Crisby, Farmer, 50
Culley, George, 117, 121, 122, 123, 127 – 167; Matthew, 121
Curtis, Kate, 97; William, 31, 37, 39, 83, 90, 91

Daniel, J., 26
Davis, John, 32; Mr.(1), 37; Mr.(2), 162
Davison, Joseph, 27, 38, 47, 49, 52, 112; Susan, 112

Derrick, William, 35, 43, 95
Dobb, Daniel, 11
Dodington, George Bubb, 118
Dolling, Mr., 74
Drake, James, 28
Duel (Duell, Dewel), A., 106; James (*Warne's uncle*), 35, 36, 41, 43, 86, 88, 106; Richard (*cousin*), 48, 77, 80; Robert (*uncle*), 32, 36, 59, 60
Dugdale, John Burleigh, 26

Effemay, Joseph, 84, 92, 110; Mr., 48
Ekins, Becky, 94; Jenny, 94; Miss, 77; Mr., 34, 104
Elliott, Emanuel, 34; John, 29, 30, 67, 106; Mrs., 25; Simon, 28

Filton, Mr., 25
Fish, Mr., 53, 54, 65, 68, 71, 73
Fisher, The Revd. Henry, 16, 69, 79, 81, 92, 103, 112, 115, 116
Florence, Betty, 85, 115, 116; Mr., 26; Mr. (*Warne's brother-in-law*), 26, 31, 36, 40, 56, 84, 85, 86, 90, 95, 103, 105; Nancy (*sister-in-law*), 31, 40, 56, 76, 84, 85, 86, 94, 95, 105; Nathaniel, 90; William, 60
Frampton, James, 8, 9, 11, 15, 28, 29, 32, 34, 35, 39, 43, 47, 49, 53 – 70, 78, 80, 81, 83, 84, 87, 90, 92 – 95, 101 – 105, 108, 119 – 121, 136
Fry, Elizabeth, 36, 37; Mr., 35; William, 56, 58, 63, 66, 96, 109
Fudge, Hannah, 54, 116; Jonathan, 65, 91; Mr., 38; Robert, 98; Thomas, 78, 80

Gallop, Richard, 55
Galton, John, 31, 54, 63, 70, 94, 100
Garland, Mr., 21, 27, 36, 39, 43, 46
Garrard (Jarrard, Jarrit, Jerrit), N., 47, 65, 79, 87
George, Mr., 57
Gold, Mr., 56
Gostelowe, Mr., 25
Gould, John, 51, 75
Graham, Mr., 145, 146, 147, 149
Grant, Alexander, 24; Gold, 25
Grose, G., 46

www.ingramcontent.com/pod-product-compliance
Lightning Source LLC
Chambersburg PA
CBHW080544110426
42813CB00006B/1201

*9 7 8 0 9 0 0 3 3 9 1 9 6 *